Guide to the Geology of Olympic National Park

Guide to the Geology o

Olympic National Park

ROWLAND W. TABOR

UNIVERSITY OF WASHINGTON PRESS

Seattle and London

Library of Congress Cataloging in Publication Data

Tabor, Rowland W
  Guide to the geology of Olympic National Park.

  Bibliography: p.
  Includes index.
    1. Geology—Washington (State)—Olympic National
Park—Guide-books.   2. Olympic National Park—
Guide-books.   I. Title.
QE176.04T3   557.97'98   74-32254
ISBN 0-295-95392-6
ISBN 0-295-95395-0 pbk.

To Kajsa
who also loves the Olympics

Preface

Now that we are becoming increasingly involved in our own man-made world, our technological brilliance, and our social convolutions, it is easy for us to forget that we are part of a continuum stretching from the elemental particles of energy and matter to the total complexity of the universe. We are born of and sustained by the planet earth, and the elements of its rocks and seas are our nearest cosmological relatives. When we can understand the earth and its history, we have come a little closer to understanding ourselves and our place in the universe. Geology is the translation of the earth's story, obscure and immensely long, into terms and time we can understand. Perhaps a little understanding of a wild mountain place like the Olympic Mountains will help us appreciate the bonds that exist between man and the universe.

Our present, albeit incomplete, understanding of Olympic geology has not been gained easily. This book sums up the hard work of many people—work that began before the turn of the century and is still going on as this is written (see "Reading and References" at the end of this book). I thank all those who have struggled with the terrain and the baffling rocks, in particular Wallace Cady, who persevered in the quest of Olympic geology longer and with greater gusto than anyone. I also thank Malcolm Clark, Howard Gower, Norman MacLeod, Kenneth Pisciotto, Weldon Rau, Parke Snavely, Richard Stewart, Martin Sorensen, Robert Tallyn, and Robert Yeats for considerable help and good discussions over the years. Also of great assistance in the field were Alan Bartow, Wyatt Gilbert, Jeanne Johnson, Robert Koeppen, and Eduardo Rodriguez.

Olympic National Park staff have been helpful throughout the work. I especially thank Mike and Paula Doherty, Glen Gallison, David Huntzinger, Bruce Moorhead, David Karraker, Louis Kirk, and Robert Kaune. Enthusiastic support at hearth and on hilltop was given by Jack and Jane Hughes.

I am grateful to all the United States Geological Survey technical staff and others who have helped transform Olympic geology into this book. Malcolm Clark, Norm MacLeod, Eduardo Rodriguez, Laurel Leone, Nancy Tamamian, Jim Pinkerton, and Kajsa Tabor helped wrestle unruly ideas and prose into subdued English. Gertrude Edmonston and Margaret Bare drafted maps. Natalie Miller, Susan Engwicht, Sara Boore, Sally Estlund, and Meade Norman drew most of the illustrations. Norm Prime, Lowell Kohnitz, and Christopher Utter provided endless photographic darkroom support. Many people typed the manuscript many times, in particular Lida Krodel and Laurel Leone.

I thank the following for permission to use previously published material: the National Geographic Society for the early topographic map of the Olympic Peninsula (fig. 3); The Geological Society of America for a photograph used for the sketch of Charles E. Weaver in figure 18 (Proceedings volume, *Geological Society of America Annual Report for 1968,* plate 28); the *Daily News* of Port Angeles for the quotation from E. B. Webster's *Fishing in the Olympics;* McGraw-Hill Book Company and Northwest Air Photos for a photograph of the Lake Cushman area in Bates McKee, *Cascadia: The Geologic Evolution of the Pacific Northwest* (1972), used as a basis for figure 32; McGraw-Hill Book Company for a photograph of a mastodon painting by Charles R. Knight displayed in the Field Museum of Natural History, Chicago, and published in *Introduction to Geology* by E. B. Branson and W. A. Tarr (2nd ed., 1941), used as a basis for figure 55. The drawing of Albert B. Reagan in figure 18 was made from a photograph in the Utah Academy of Sciences, Arts, and Letters *Proceedings* 16 (1939): 4.

Contents

Illustrations

Maps

Figures

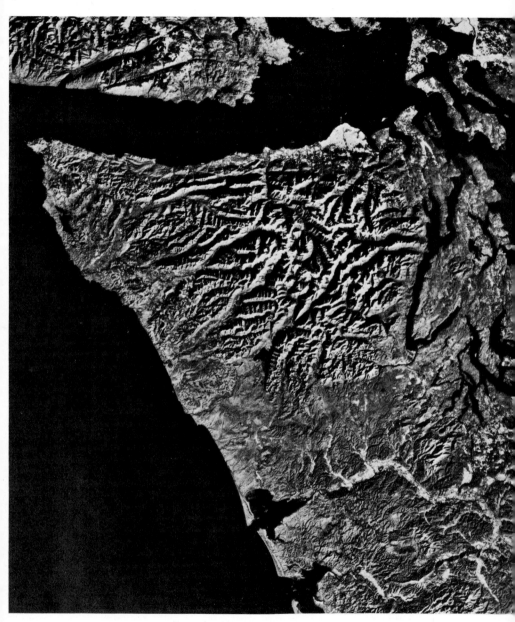

Olympic Peninsula as viewed from the Earth Resources and Technology Satellite about 570 miles above the earth. Photo-mosaic by U.S. Geological Survey

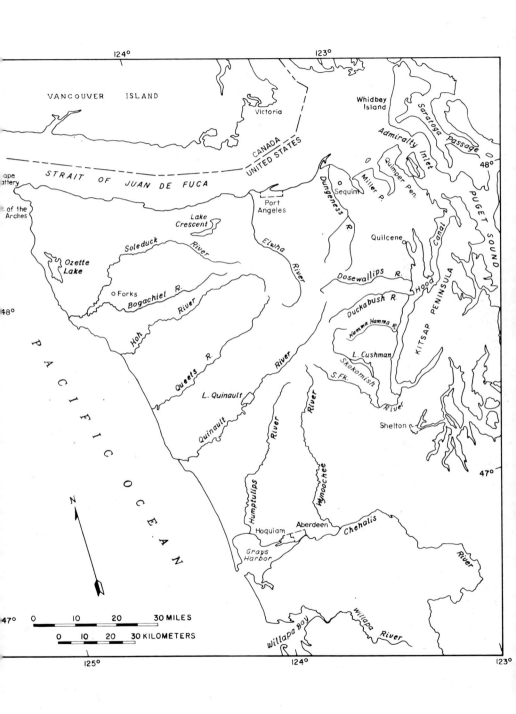

Guide to the Geology of Olympic National Park

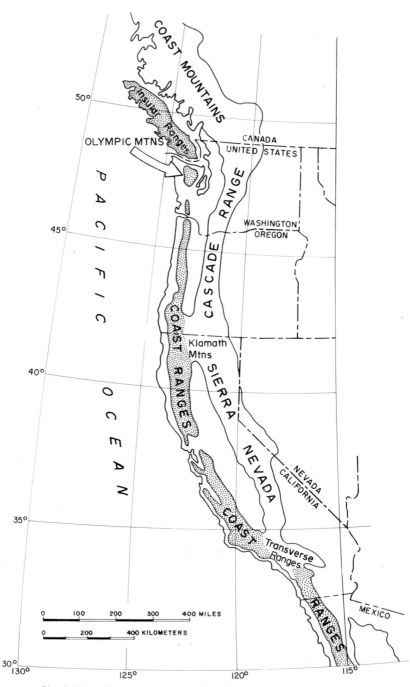

Fig. 1. Mountain ranges along the Pacific coast

Introduction

The Unique Mountains

Along the west coast of North America, from Mexico to southern Canada, are mountain ranges of diverse character collectively called the Coast Ranges (fig. 1). The Olympic Mountains, at the extreme northwest corner of the conterminous United States, are a unique part of these ranges. Even though they are closely related in rock composition to the Coast Ranges of Oregon, they are separated from them by the broad lowland of the Chehalis River and are considerably higher and more rugged. They have some scenery in common with the Insular Ranges of Vancouver Island in Canada but are quite different geologically.

Viewed from above, the Olympics seem to present a disorganized, circular array of jagged peaks above a deep, forested labyrinth of canyons (fig. 2); but the dominant design is controlled by eleven major rivers radiating from the mountains like the spokes of a wheel. This circular spread and radial river plan (see pages xiv-xv) show that the Olympics developed as a separate uplift, not as a part of a long, coastal mountain chain. They comprise a mountain massif in themselves. Between the major rivers in the core of the range are extensive tracts of alpine and subalpine terrain: flowered meadows, barren rocky expanses, and glacial ice.

The major rivers—the Skokomish, Hamma Hamma, Duckabush, Dosewallips, Dungeness, Elwha, Soleduck, Bogachiel, Hoh, Queets, Quinault, Humptulips, and Wynoochee—carry a tremendous volume of water because the high Olympics, intercepting Pacific storms, receive more rain and snow than any other place in the conterminous states. In the coastal area, precipitation averages about 140 inches per year and in the high mountains it may approach 200 inches; but on the northeast side of the peninsula, in the lee of the mountains, rainfall decreases rapidly to less than 20 inches in the Sequim area. Most rain and snow falls in the winter months, but the rivers are fed around the year by melting winter snow and glaciers.

Fig. 2. Deep-forested canyons of the Olympic Mountains. View northeast down Grand Creek

Early Exploration: The Geographic Map

The Olympics lay obscure and remote long after the rest of the western United States was well explored and mapped. Although there was some settlement around the periphery of the Olympic Peninsula, maps of the mountain region made as late as 1890 were blank in the center.

Official exploration of the mountains got under way in 1885 when a party led by Lieutenant Joseph O'Neil of the U.S. Army reached the high country of the Hurricane Ridge area (see relief map) and explored to the south. O'Neil returned in the summer of 1890 to construct a mule trail up the North Fork of the Skokomish and down the Quinault. While accomplishing this remarkable task, he and his exploring parties also traveled the South Fork of the Skokomish, the Humptulips, the Wishkah, the Satsop, the Wynoochee, the North Fork of the Quinault, the Upper Elwha, and the Queets!

Lieutenant O'Neil's monumental but little-publicized exploration was outshone by the efforts of another party, the *Press* Exploring Expedition, sponsored, and well publicized, by a Seattle newspaper. The *Press* party, launched with great journalistic fanfare, barely made it across the mysterious range via the Elwha and Quinault rivers. Robert L. Wood tells the almost comical story of this expedition in his delightful book, *Across the Olympic Mountains: The Press Expedition, 1889–90* (see "Reading and References").

After these explorations, peaks and rivers finally began to be recorded on maps (fig. 3). Wood has neatly summed up the story of Olympic exploration and the evolution of Olympic National Park in another book, *Trail Country: Olympic National Park* (see "Reading and References").

Despite increasing travel and geographic exploration in the back country, little has been known about the geology of the mountains until recent times. The geology has remained obscure largely because of the mountains' steepness, heavy vegetation, and inaccessibility. Olympic rocks are also difficult to understand, and only recently have geologists had the concepts and tools to work with such rocks.

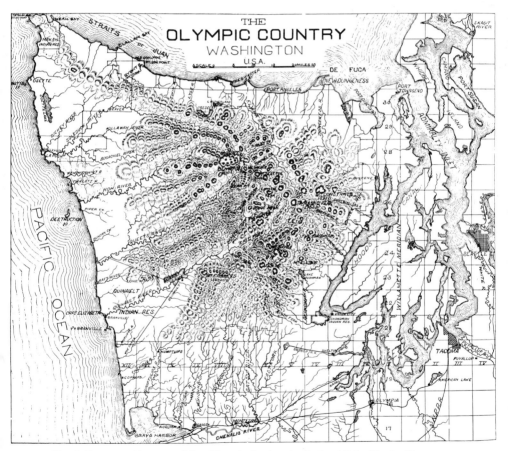

Fig. 3. Topographic map of the Olympic Peninsula, first published in the National Geographic in 1896. The topography of the central portion is based on the explorations of Mr. S. C. Gilman. Courtesy of the National Geographic Society

How the Geologist Explores: The Geologic Map

The organization of geologic knowledge begins with the geologic map, which shows the distribution of different kinds of rock. To make a geologic map of an area like Olympic National Park, which has only 128 miles of road in 1,400 square miles of rugged mountainous terrain, the geologist must, first of all, walk. He must hike the trails, climb the trailless ridges, and wade down the log-choked creeks to find *outcrops,* that is, exposures of bedrock. At times a pack train may carry his duffle, or a helicopter may whisk him from one range to another; but to see a significant amount of rock, he must travel on the ground, as close to the rock as his legs will carry him (fig. 4).

At each outcrop, the geologist notes the kind of rock, its characteristics, and its geometry, and marks its position on the map. For instance, he may find an outcrop of sandstone. He notes the size of its grains, its color, and the characteristics of the bedding. He measures the position of the bedding, which is especially important if the rock has been folded. He looks for fossils and collects a specimen of the rock to take back to the laboratory where he can study it with a microscope or analyze its chemical components.

As the number of notes on the map grows, he develops a hypothesis to explain what he sees: the green sandstone is older than the

Fig. 4. Geologist at work on Mount Carrie. Dodger Point, Ludden Peak, and Mount Ferry are in the middle background; The Needles, Mount Deception, and Mount Constance on the left skyline; and The Brothers, Mount Anderson, and Chimney Peak on the right skyline

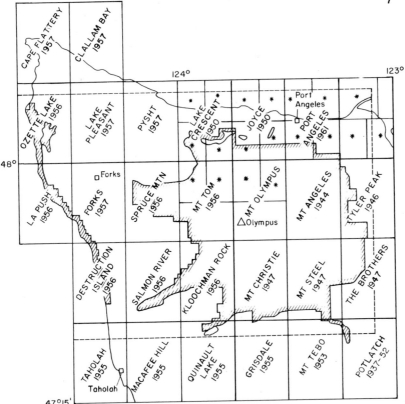

Fig. 5. Index to topographic quadrangle maps of the Olympic Mountains. Named maps at 1:62,500 scale. Asterisk marks location of maps available at 1:24,000 scale. Dashed line is the boundary of the map of Olympic National Park and vicinity at 1:125,000 scale, available with or without shaded relief

black shale; both have been folded and eroded before a younger conglomerate was deposited on top of them. To test this hypothesis, he makes more observations, then modifies or rejects it accordingly. The pattern of rock distribution on his map becomes a geologic map (see fig. 16).

In his head and in his notes, the geologist develops a geologic history. However, the story is not always simple, and there may be several reasonable versions of how things happened. The version of Olympic geology presented here is only one of several possibilities, but it seems to be the best at present.

Using This Guide

This book is divided into two parts. The first part summarizes the geology of the Olympic Mountains and relates their geologic history.

The second part guides the reader to places where he can observe geologic features. Many of the more fundamental geologic principles are explained in the second part, on the outcrop, so to speak. The geologic story is keyed to these points of interest, allowing readers to expand their knowledge of topics or terms. The glossary and index should help readers find their way through a welter of geologic terms and concepts.

The guide to points of interest (Part II) is organized around drainages—the eleven major rivers and a few smaller ones. This may prove a little awkward for the traveler crossing from one drainage to another, but most visitors will find the geologic notes sequentially arranged along their route, at least as they go into the mountains.

Here I must add a note of caution. Olympic National Park is a special place to enjoy rocks. Find them, admire them, learn from them; but do not collect them. Collecting is taboo without a special permit; if it were not, an outcrop of unusual rock might virtually disappear under the onslaught of rock hounds and geologists.

The locations of points of interest are given on the shaded-relief map. This map, at a scale of about 1:138,750 (that is, 1 inch or foot on the map equals 138,750 inches or feet on the ground, or about 1 inch to 2.2 miles) is adequate for route-finding on most trails; but to reach off-trail points of interest, the serious hiker should carry larger-scale maps such as the standard quadrangle maps at 1:62,500 (about 1 inch to the mile) or the larger, less convenient maps at 1:24,000 (about 2 inches to the mile) (fig. 5). These maps are all available by mail from the United States Geological Survey, Federal Center, Denver, Colorado 80225 or Washington, D.C. 20244 at $.75 each. Olympic National Park Museum and a few private firms also carry some quadrangle maps.

PART I
Olympic Geology

Hand sample   Under the microscope

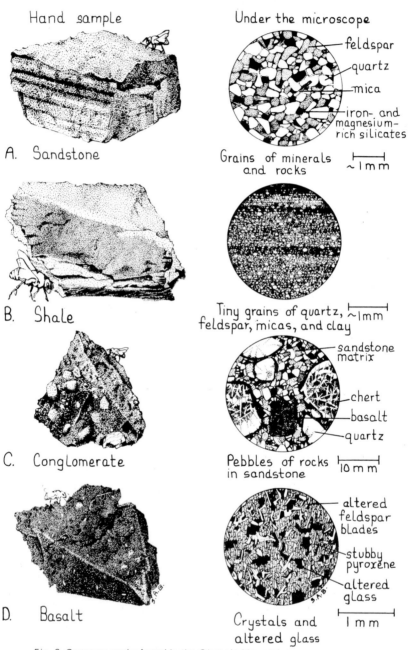

A. Sandstone   Grains of minerals   ⊢———⊣
                               and rocks   ~1mm

B. Shale   Tiny grains of quartz, ⊢———⊣
                               feldspar, micas, and clay   ~1mm

C. Conglomerate   Pebbles of rocks ⊢———⊣
                               in sandstone   10 mm

D. Basalt   Crystals and   ⊢———⊣
                               altered glass   1 mm

Fig. 6. Common rocks found in the Olympic Mountains

The Making of the Rocks

*". . . these mountains bear every indication of being of very
recent formation and I fully believe they are."*

Lieutenant Joseph P. O'Neil
14th Infantry, 1890

Rocks of the Olympics

Fortunately for the nongeologist, the basic rock vocabulary of the
Olympic Mountains is simple. The visitor can easily recognize the
three most abundant rocks: *sandstone, shale,* and *basalt.* Most
Olympic rocks were born in the deep waters of the Pacific Ocean,
where sand and mud from the land accumulated, bed by bed, into
thick, nearly horizontal layers. Adding to this pile of rock were flow
after flow of basalt lava that issued from cracks or broad volcanoes on
the ocean bottom. Because of this origin, almost all the rocks of the
Olympics reflect the assaults and caresses of ocean currents, the
buoyancy of water, and the cold and pressure of ocean deeps.

The build-up of these sediments and lavas, represented now by
rocks in the Olympic Mountains, spanned a period of some 40 million years. The oldest rocks are about 55 million years old, the
youngest about 15 (see fig. 12). By geologic standards, these are indeed young mountains made of young rocks.

Sandstone. Sandstone forms from sand—small grains of mineral
and rock eroded from pre-existing rocks (fig. 6). Usually the grains
can be seen easily: glassy grains of *quartz,* white grains of *feldspar,*
shiny black or clear flakes of *mica,* and green or black grains of rock
or iron- and magnesium-rich silicate minerals.

It is obvious that sand grains are carried to the sea by streams and
rivers, but for a long time geologists puzzled over how sand could
travel through quiet water into the deep ocean. The answer came from
careful study of deep-ocean sands and laboratory experiments. Features in sandstone beds indicate that the sand was swept into deep
water in a slurry of sand, silt, and water that flowed along the ocean
floor. Such slurries, called *density currents,* maintain their identity and
flow downhill under the ocean just as streams of water on land maintain their identity and flow under a sea of less dense air. Some sandstone beds deposited by density currents are so uniform and struc-
tureless that they look as though they were delivered to their resting

11

Fig. 7. Thick beds of sandstone near Windfall Peak. Hurricane Ridge is in background, to the northeast

places all at once. Other beds are graded, the coarsest sand at the bottom, the finest at the top. Gentle ocean-bottom currents may pick up and redeposit upper layers of the sand, producing still other varieties of bedding. In addition, marine worms and other boring creatures may leave their tracks on and in the beds. A variety of deep-water events are recorded in these sand beds and can be recognized millions of years later (fig. 7, and notes 111, 123).

Shale. Shale is rock made from mud. Olympic shales are commonly made up of clay minerals, quartz, feldspar, and micas (fig. 6), although the particles are usually too fine to be seen easily. In contrast to coarser sand, these fine particles were swept into deep ocean water and suspended by ocean currents. The particles settle out of suspension slowly and can drift far from shore before deposition. Whether or not this fine material becomes a shale bed—and if so, how thick it is—commonly depends on how much material rains down and how much time passes between slurries of sand. In some places the sand flowed out at very regular intervals, giving rise to a rhythmic alternation of shale and sandstone beds (fig. 8, and notes 55, 135).

Fig. 8. Rhythmically bedded sandstone and shale on Blue Mountain. Beds that were originally deposited horizontally have been deformed and now stand vertically

Transformation of the mud and sand into shale and sandstone takes millions of years of deep burial. The weight of overlying sediments squeezes out water and presses the grains tightly against each other. Minerals such as calcite, limonite, and quartz precipitate from water between the grains to cement them together. Eventually, the sediments become rocks.

Conglomerate and Sedimentary Breccia. Occasionally while Olympic sands and muds were accumulating on the ocean bottom, a particularly thick slurry would carry rounded pebbles and cobbles originally from the continent into the deep ocean. Time and pressure have bound these gravels into *conglomerate* (fig. 6, and notes 55, 99, 108).

At other times, as a slurry rushed along the ocean bottom, it ripped off chips of partly consolidated mud and deposited them nearby. *Shale-chip* and *slate-chip breccias,* as these deposits of angular fragments are called, make striking spotted outcrops in many areas of the Olympics (fig. 9).

Basalt and Its Associated Rocks. Most Olympic basalt is black or dark green, although weathered surfaces may be red or yellowish owing to chemical changes in the iron-bearing minerals. Basalt is hard, dense rock made up mostly of tiny crystals readily seen only with a microscope (fig. 6). Basalt forms from molten rock, called *lava,* that has erupted on the earth's surface. Bubbles of gas in the molten lava are commonly trapped in the rock, leaving globular holes, or vesicles. Later, these vesicles are commonly filled with zeolite minerals (note 66), giving the rock a polka-dot appearance.

Sometime after sediments began accumulating in the ocean, molten rock poured out on the ocean floor. We know that Olympic basalt had this watery beginning because it is commonly made up of globular masses called *pillows,* which form primarily when hot lava erupts under water (notes 25, 60).

The basalt lavas piled up to incredible thicknesses not only as pillows but also as piles of broken rubble called *volcanic breccia* (notes 60, 63). The actual thickness of the pile is not known, but on the east side of the Olympics, the lava and breccia beds, now tilted on end, measure more than twelve miles thick. Some thickening results from folding, and some parts of the pile may be repeated on account of faulting (note 26). Even so, the pile is enormous by man's standards.

The lavas piled up so high on the ocean floor in some places that they protruded above the surface as islands, even though the floor was probably sinking under the great load. Lava flows erupting on these islands were not chilled as quickly as those in the water, so they did not form pillows. Instead, as they cooled and shrank they developed cracks that broke the lava into polygonal columns (fig. 10). *Columnar joints* are typical of lavas erupted on land, and they may be seen in roadcuts along Hood Canal.

Also within the basalt pile and in some of the sedimentary rocks are thin layers of basaltic rock that never reached the ocean floor but solidified under a blanket of rock or sediment. These *dikes* and *sills* (fig. 11) cooled slowly, and crystals grew larger in them than in the quickly chilled lavas. If the crystals are just large enough to see, we call the rock *diabase;* if they are coarse enough to rest a little finger on, *gabbro* (notes 2, 46).

Fig. 9. Slate chips in sandstone (slate-chip breccia) below Geri-Freki Glacier. The smooth outcrop surface and the faint grooves (striations) that trend downslope are evidence of recent glacial action

Fig. 10. Columnar joints in basalt, formed
when cooling lava shrinks in the direction
of the arrows

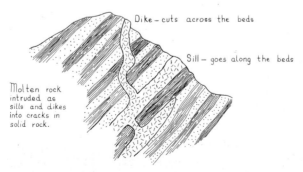

Fig. 11. Dikes and sills

Interbedded with the pillow lava is red limestone rich in shells of one-celled animals (Foraminifera) that thrived in deep water (notes 2, 60). Also abundant are the submicroscopic plates of floating animals called plankton. Some limestones appear to be formed entirely of these plates.

Of special significance to technological man are the manganese- and copper-bearing minerals closely associated with the red limestones. Apparently, the metal-bearing minerals formed when volcanic gases and solutions from the lavas reacted with sea water during or shortly after volcanic eruption. Small deposits of copper and manganese were prospected and mined at various times in the early twentieth century around the periphery of Olympic National Park. There was much activity at the Crescent Mine near Lake Crescent, the Elkhorn Mine on the south side of Mount Constance, the Tubal Cain Mine northeast of Buckhorn Mountain, and the Black and White Mine near Mount Cruiser (see notes 4, 5, 10, 23). The occurrences of ore minerals in the Olympics are now more curious than commercial, and man has gone elsewhere for his supplies of manganese and copper.

The entire pile of basalt and its associated rocks has a primitive chemical composition. We can therefore conclude that the lava is derived from material deep in the earth that has never been subjected to the chemical and mechanical sorting processes of the earth's surface. Olympic lavas have much in common with those that make up most of the Pacific Ocean floor and numerous undersea volcanoes active today.

Rocks and Time

Contemplation of earth history is always hindered by the mind-boggling dimensions of geologic time. We can talk about a million years or even a billion years, but we can hardly imagine the countless small events that fill such expanses of time. And it is just such small events—the settling of a sand grain to the ocean bottom, the tumbling of an unsteady pebble into a creek, the death of a small snail—that add up to geologic change.

If we travel down the Elwha River from Lake Mills to Lake Aldwell, we will have traversed in about 5 miles a section of rock—basalt, shale, and sandstone—representing about 10 to 15 million years of accumulation. If we were to walk through time at this same rate, a few paces more would encompass all of mankind's recorded history, and a slight shuffle of an inch or so would take in one person's lifetime (fig. 12). To pace out the whole of the earth's history since its beginning about 4.5 billion years ago, we would have to hike about 1,500 miles.

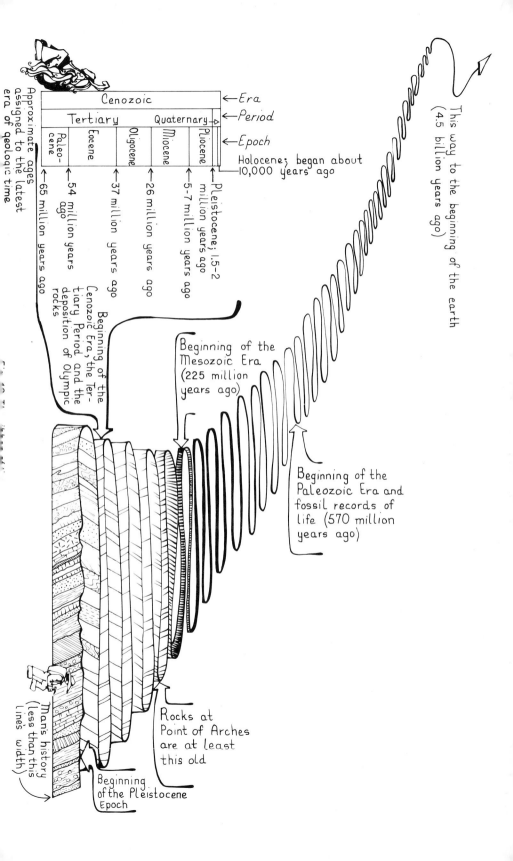

This way to the beginning of the earth (4.5 billion years ago)

Approximate ages assigned to the latest era of geologic time

Cenozoic					←Era
Tertiary			Quaternary		←Period
Paleo-cene	Eocene	Oligocene	Miocene	Pliocene	←Epoch

Holocene; began about 10,000 years ago

←65 million years ago

←54 million years ago

←37 million years ago

←26 million years ago

←5-7 million years ago

←Pleistocene; 1.5-2 million years ago

Beginning of the Cenozoic Era, the Tertiary Period and the deposition of Olympic rocks

Beginning of the Mesozoic Era (225 million years ago)

Beginning of the Paleozoic Era and fossil records of life (570 million years ago)

Rocks at Point of Arches are at least this old

Man's history (less than this lines' width)

Beginning of the Pleistocene Epoch

The Oldest Rocks. Geologists are wont to ask upon what all the sedimentary and volcanic rocks of the Olympics were deposited, a question not unlike that of the Greeks who wondered where Atlas stood as he held up the earth. The answer is not entirely clear, but most of the Olympic rocks may have been laid down directly on basaltic crust under the Pacific Ocean. Some rocks, however, were deposited on the eroded surface of older igneous and sedimentary rocks —perhaps a submerged piece of the North American continent of 70 million years ago. Just south of Point of Arches, modern sea cliffs and sea stacks of old gabbro, basalt, and sandstone may represent this piece of continent.*

Events Deep in the Earth

Soon after the basalt, sandstone, and shale accumulated, and probably even while they were accumulating, slow movements of the earth's crust began to squeeze and break the rocks. The bedded sandstone and shale were folded and then folded again. In thick masses, bedding was broken, stretched apart, and squeezed along in softer sediments (notes 35, 105, 125, 138). The thick, strong basalts resisted folding to some extent and they broke in many places. Thin flows of basalt that trailed out into the pile of sediments were broken off in whole chunks and squeezed along in a mush of shale and soft sandstone. Thick masses of hard sandstone were also mixed in—like nuts in pudding. At times when squeezing waned, solutions rich in silica deposited quartz in cracks. The rocks were again folded, and when folding ceased, new cracks formed and once more filled with quartz. Quartz-veined rocks are conspicuous in some areas (notes 100, 172, and fig. 13).

During this disruption many of the rocks were becoming harder. The heat and pressure deep in the earth made the old minerals in the sandstone and shale and basalt react with each other and with solutions in the rocks. This *metamorphism* produced new minerals (notes 103, 136). The squeezing elongated sand grains, made pebbles into rods, and filled the shales with minutely spaced cracks. As individual sand grains were smeared out, the sandstone developed streaky lay-

* Some of the old rocks at Point of Arches (fig. 16) are at least 144 million years old, that is, at least 80 million years older than the oldest rocks in the Olympic Mountains (fig. 12). Cliffs at Point of Arches, reached via Shi Shi beach on the Makah Indian Reservation, reveal Eocene conglomerates and breccias filled with large boulders eroded from the old continent. The intrepid explorer can reach the old gabbro—a rock of white (feldspar) and black (hornblende) spots—by scrambling along the cliffs south of the point at low tide. To approach the gabbro from the south, drive to the beach north of the Ozette River and hike north at very low tide.

Fig. 13. Light-colored quartz veins in sandstone west of Gray Wolf Pass

ering, and some of it became *semischist* (fig. 14, and note 105). Shale became *slate* and *phyllite* (figs. 14, 15, and notes 64, 165), and basalt recrystallized to *greenstone* which, although it is greener, still looks like basalt (notes 12, 110).

Theories for the Origin of Olympic Structure

Looking at and recognizing the rocks of an area and how they formed is only the first step in interpreting the geologic story. We would like to see how all the different rocks are related to each other. The pattern of relationships, portrayed by a geologic map (fig. 16), helps decipher the geologic structure; from this we can begin to reconstruct the geologic history. The ultimate goal is to fit the local geologic history into a total earth history.

The Earliest Ideas. Some of the observers in the Olympics thought that the bedrock of the mountainous core, sometimes called the inner Olympics (fig. 17), must be composed of gneiss and granite, rocks known to be common to the cores of many mountain ranges. The

Hand sample

visible mica flakes

Under the microscope

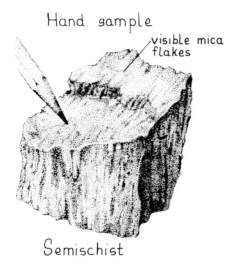

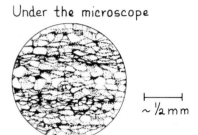

~ ½ mm

Broken grains of quartz and feldspar in a web of micas

Semischist

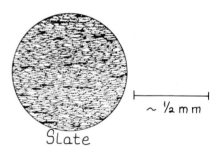

~ ½ mm

Slate

Barely recognizable grains of quartz, feldspar, and micas in parallel planes

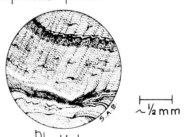

~½ mm

Slate

Phyllite looks much the same as slate but is shinier

Phyllite

Easily recognizable grains of quartz, feldspar, and micas in parallel planes which are commonly folded

Fig. 14. Metamorphism changes sandstone to semischist and shale to slate and phyllite (cf. fig. 6)

Fig. 15. Folded slate and phyllite near Mount Norton

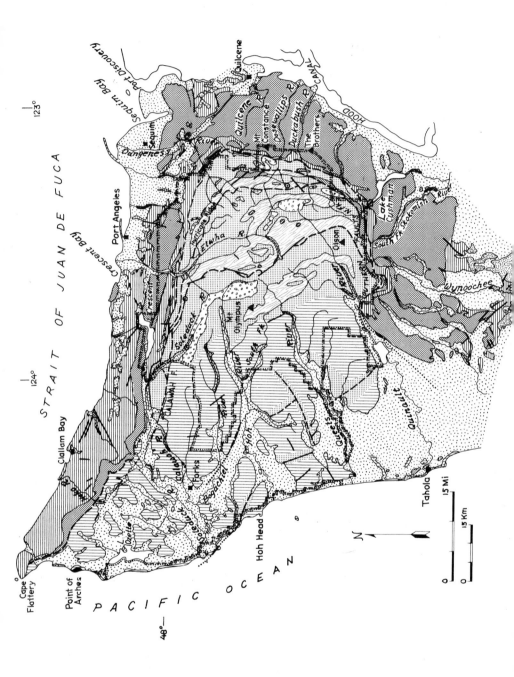

STRAIT OF JUAN DE FUCA

PACIFIC OCEAN

Cape Flattery

Point of Arches

Clallam Bay

Port Angeles

Crescent Bay

Sequim Bay

Port Discovery

Quilcene

HOOD CANAL

Dungeness

Sequim

Elwha R.

Quilcene R.

Mt. Constance

Dosewallips R.

Duckabush R.

The Brothers

Lake Cushman

South Fk. Skokomish River

Wynoochee R.

Crescent R.

Sol Duck R.

L. Ozette

Calawah R.

CALAWAH F.

Bogachiel R.

Forks

Forks R.

Hoh R.

Hoh Head

Hoh River

Queets R.

Quinault R.

Taholah

Olympus Mt.

123°

124°

48°

N

15 Mi

15 Km

0   15 Km

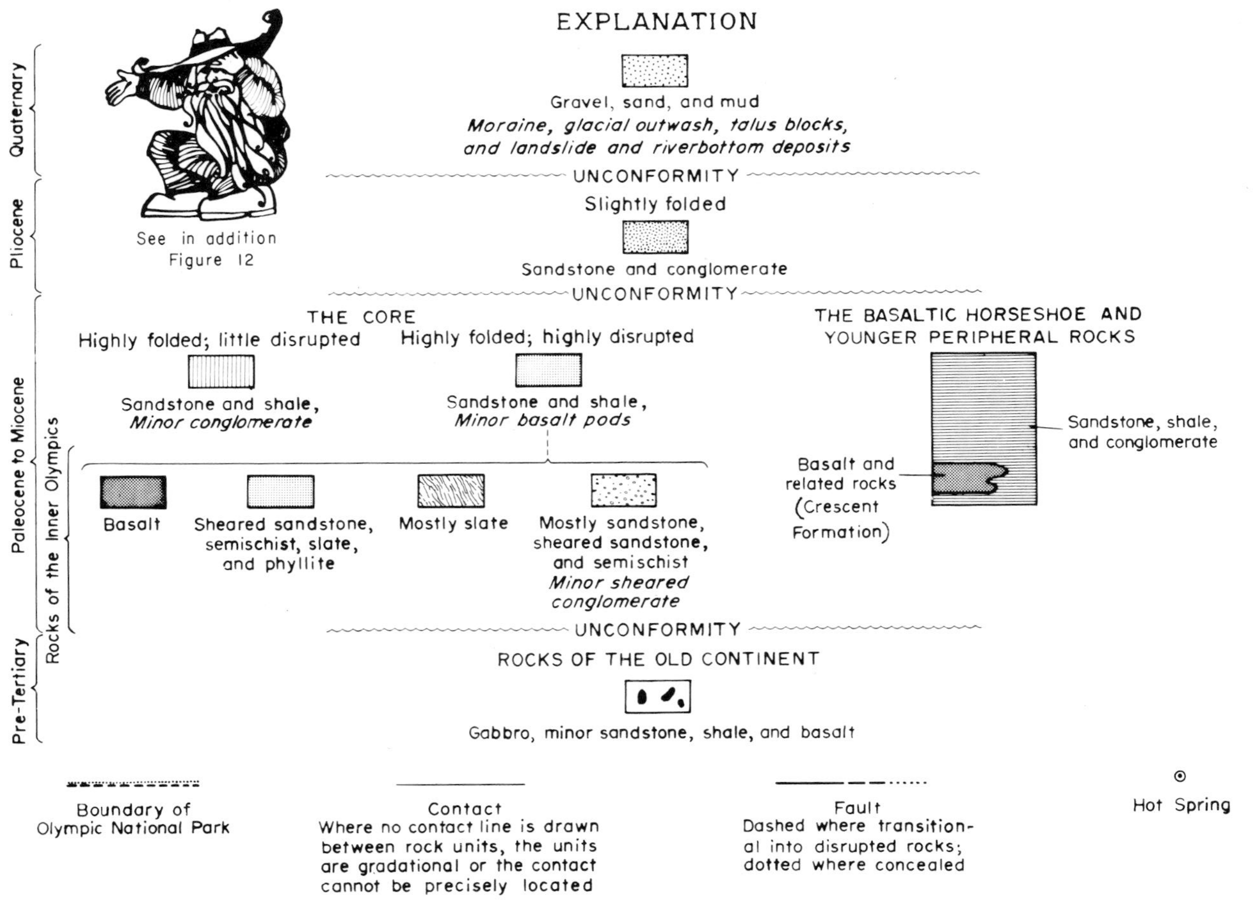

EXPLANATION

Gravel, sand, and mud
Moraine, glacial outwash, talus blocks, and landslide and riverbottom deposits

UNCONFORMITY

Slightly folded

Sandstone and conglomerate

UNCONFORMITY

THE BASALTIC HORSESHOE AND YOUNGER PERIPHERAL ROCKS

Sandstone, shale, and conglomerate

Basalt and related rocks (Crescent Formation)

THE CORE

Highly folded; little disrupted | Highly folded; highly disrupted

Sandstone and shale, *Minor conglomerate*

Sheared sandstone, semischist, slate, and phyllite

Basalt

Sandstone and shale, *Minor basalt pods*

Mostly slate

Mostly sandstone, sheared sandstone, and semischist *Minor sheared conglomerate*

ROCKS OF THE OLD CONTINENT

UNCONFORMITY

Gabbro, minor sandstone, shale, and basalt

See in addition Figure 12

Rocks of the Inner Olympics

Quaternary | Pliocene | Paleocene to Miocene | Pre-Tertiary

Boundary of Olympic National Park

Contact
Where no contact line is drawn between rock units, the units are gradational or the contact cannot be precisely located

Fault
Dashed where transitional into disrupted rocks; dotted where concealed

⊙ Hot Spring

Fig. 16. Simplified geologic map of the Olympic Peninsula

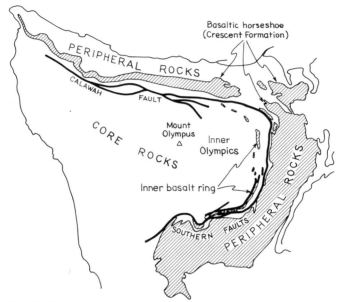

Fig. 17. Predominant geologic terranes of the Olympic Peninsula

early travelers were fooled, no doubt, by the gneiss and granite occurring as pebbles and boulders in the gravels of the major streams on
the north and east flanks of the range. We now know that these exotic
rocks do not occur as bedrock in the Olympics but instead were carried and dumped there by the Cordilleran ice sheet as it flowed
against and around the mountains.

   One of the earliest expeditions to penetrate the mountainous interior of the Olympics—the *Press* party of 1889–90—correctly determined that at least some of the core was composed of slate and sandstone; but their scientific observations were not taken very seriously,
and they lost all of their specimens when their raft capsized on the
Quinault River.

   By the early 1900s geologists were hammering on rocks around the
periphery of the range and venturing up trails along the major drainages. Among these early workers was Albert B. Reagan, an Indian
agent and writer with an insatiable curiosity (fig. 18). Reagan published not only on the geology of the area but also on the flora, the
fauna, and, of course, the Indians. He collected numerous Indian legends, a few of which explained, with some basis of truth, geologic phe-
nomena in the Olympics (notes 112, 134).

   A more modern approach to the geology was taken by Charles
Weaver, a geologist-paleontologist from the University of Washington

Fig. 18. Albert B. Reagan, left, and Charles E. Weaver, right, early workers in the Olympics. Sketch of Reagan is based on a photograph in the Utah Academy of Sciences, Arts, and Letters Proceedings 16 *(1939):4; sketch of Weaver made from a photograph first published in the* Geological Society of America Annual Report for 1968

(fig. 18). Weaver studied an incredible amount of Tertiary sedimentary rock along the western coast of the United States. In the course of this monumental study, he examined most of the rocks on the north side of the Olympic Peninsula and outlined the horseshoe pattern of basalt (figs. 16, 17). Today the basalt is called the Crescent Formation (named for exposures near Port Crescent—now Crescent Bay—not for its crescent outcrop pattern). In 1937, Weaver proposed that the Olympics were part of a giant plunging arched fold, the top of which had been eroded off. If layers of sediment are folded up in the middle, they form an *anticline;* and if the top of the anticline is eroded off, the older layers will then appear at the surface in the center of the fold (fig. 19).

Weaver believed that the rocks of the Olympic core were older and therefore represented the center of the anticline. This was a reasonable guess, since most of the core rocks looked older; they were more highly deformed and certainly harder than the peripheral rocks. He could not prove that all the core was older, however, for although he and others had found Eocene and younger fossils in the peripheral rocks, they found no fossils in the mountainous part of the core.

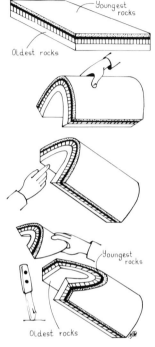

Fig. 19. Development
of a plunging anticline

Meanwhile, oil company geologists, prospecting along the western part of the peninsula, found fossil Foraminifera, or "forams" as they are fondly called (fig. 20). The location of these fossils indicated that at least some of the core rocks encircled by the arms of the basaltic horseshoe were younger than the basalt, thus casting some doubt on Weaver's theory. In spite of this, his anticline hypothesis has persisted into modern times.

Development of a Geologic Map

"The great tragedy of Science—the slaying of a beautiful hypothesis by an ugly fact."
T. H. Huxley, Presidential Address,
British Association for the Advancement of Science, 1870

By the 1940s geologists from the United States Geological Survey and geology students from the University of Washington were beginning to fill in more of the geologic map of the Olympics. From 1938 to 1941 an ambitious mapping project was undertaken by a Geological Survey team led by Charles Park, Jr. The government sent him into the Olympics to assess the supplies of manganese, a metal then difficult for the

United States to obtain because of world conflict. As they searched for manganese minerals, Park and his team of young geologists also ferreted out almost all the outcrops of basalt and greatly improved the geologic map of most of the basaltic horseshoe, outlined earlier by Weaver. Although Park was interested primarily in the manganese and its origin, he recognized that the simple anticline theory was not quite right. First, clam shells of probable Oligocene age had been found near Mount Appleton, well within the mountainous core. This further confirmed that at least some rocks in the core were younger than the basalt flanks of the so-called anticline. Second, Park recognized the complexity of faulting in the core rocks and suspected similar complexities in the basaltic horseshoe, especially on its east side. He thought that the core sandstone and slates, as well as the basalts, appeared much thicker than they originally were because they had been cut and doubled up by faulting (fig. 21).

Not long after Park's work, an intense student from the University of Washington, Wilbert Danner, began mapping in the Olympics. Danner had fallen in love with the Olympics as a Boy Scout. In 1955 he summarized Olympic geology in a booklet that became the definitive work on Olympic geology for geologists and nongeologists alike (see "Reading and References"). Danner still believed the mountains were an anticlinal feature, with older rocks in the core; but he reported the possibility that younger rocks were there too. By the 1950s more students from the University of Washington were thrashing through the underbrush on the east and south sides of the mountains, and a more coherent picture of Olympic geology was evolving.

By that time it was clear that the inner Olympics are partly ringed by Eocene basalt, the Crescent Formation, forming an aggregate of ridges called the basaltic horseshoe. The rocks outside the basaltic horseshoe and overlying it are mostly sandstone, shale, and conglomerate. Although folded and faulted, they are nowhere as disturbed as

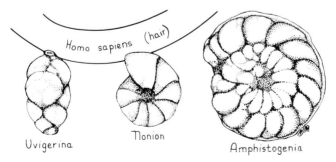

Fig. 20. Foraminifera shells

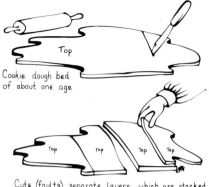

Cookie dough bed
of about one age

Cuts (faults) separate layers, which are stacked

Fig. 21. Cutting, stacking, and folding cookie dough "bed"

Faulted slices are on top of one another as in a stack of rocks but are all of one age

the rocks within the horseshoe. Within the core is a second partial, broken ring of basalt, called the inner basalt ring (figs. 16, 17 and notes 18, 46).

Meanwhile, another Geological Survey team consisting of Robert Brown, Jr., Howard Gower, and Parke Snavely, Jr., was wading up and down the creeks, crawling through the brush, and scrambling over the northern foothills of the Olympics. Since the early days of Albert Reagan, many new roads had been cut through the thick Olympic forest, greatly improving the geologists' chance of reaching outcrops. The team was able to gather many new data, plot them on modern topographic maps, fit them together with earlier pieces of information, and finally produce a coherent history of the rocks along the northern periphery of the mountains.

One significant discovery of the U.S. Geological Survey mappers was the Calawah fault zone (notes 81, 83). This zone of highly broken rocks and some smaller faults separate the relatively simple rock structures of the north periphery from the more complex ones of the core. Recognition of the fault zone suggested the possibility that the rocks reportedly younger than the Eocene basalt were brought into the core by fault movement.

Another important piece of the puzzle was contributed by Richard

Stewart, a zealous geology student who mapped the difficult tree-covered terrain of the western core. Stewart showed that these rocks too were complexly disturbed, although less so than rocks in the core to the east. He also carefully analyzed the sedimentary features of the rocks to show that the sandstones and shales were probably deposited by density currents near the continent in a thick wedge of sediment called a *submarine fan.*

Credit for the geologic mapping of the Olympics should go also to the paleontologists who worked with fossils collected by the mappers. In particular, Weldon Rau, a paleontologist for the state of Washington and a field mapper in his own right, studied hundreds of samples of shale to identify Foraminifera shells. The ubiquity of Foraminifera in marine rocks and their rapid evolutionary changes help to distinguish beds of different ages, and the study of these small shells has been a tremendous aid to the geologists in trying to untangle jumbled beds.

One of the young geologists with Charles Park's team of manganese mappers was a perambulatory and persistent New Englander named Wallace Cady. Like Danner, Cady had started in the Olympics as a Boy Scout. His love of the mountains brought him back not only after graduate school to accompany Park in his explorations but again in the 1960s to head a third Geological Survey team that began systematic mapping of the northeastern part of the range. Our work, later expanded to all the enigmatic rocks of the Olympic core, forms the basis of ideas set forth in this book. We were greatly aided not only by all the work that preceded us, but also by new tools: new concepts in sedimentary processes, a new and exciting theory of earth evolution —and the helicopter!

Ideas of Today. Mapping the core rocks has clarified several problems. Except for the ancient rocks at Point of Arches, no rocks older than those of the Eocene basaltic horseshoe and some minor sandstones and shales underlying it have yet been found in the core. A fair number of Foraminifera in red limestone from scattered places in the core appear to be no older than Eocene. One fossil clam shell and several snail shells from near The Needles have been dated by paleontologist Warren Addicott (fig. 22). They are definitely Eocene and

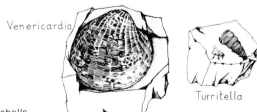

Venericardia

Turritella

Fig. 22. Clam and snail shells

probably late Eocene. The scarcity of fossils in the core rocks is somewhat mysterious, but metamorphism and deformation probably have destroyed the fragile shells. Less metamorphosed rocks to the west contain many more microfossils, and as earlier workers surmised, some are as young as Miocene.

We now know that the core is made up of arcuate belts of rock roughly concentric with the basaltic horseshoe and separated from it and from each other by faults. The rocks of these belts are highly folded, broken, and metamorphosed. The most intense disruption and the greatest metamorphic changes are near the center of the bend in the horseshoe (notes 103, 105, 136).

A fault or series of faults bounds the disrupted core rocks on the south in much the same way the Calawah and other faults bound them on the north (figs. 16, 17, and notes 9, 18).

In spite of the folding and breaking of the core rocks, most of the original tops of the beds (that is, the direction that was *up* when the sediments were deposited on the ocean floor—see notes 63, 111, 123) face away from the core of the range. The pattern still looks like a simple anticline, but it cannot be, for the core rocks are not older than the flanks (fig. 16). Charles Park appears to have been correct when he suggested that the thick mass of core rocks was thickened by fault slicing. The whole pile can be likened to a sheet of cookie dough, cut up into pieces (by faults), then piled up in a stack (by thrusting), turned on edge, and bent (fig. 21).

To find a plausible explanation for this extreme rock deformation and Olympic structure in general, we must digress from Olympic rocks and examine some of the newer ideas about earth evolution.

Forces in the Earth: Plate Tectonics

Geologists have long sought an explanation for how sedimentary rocks derived from flat-lying sediments in the sea could be so folded and compressed and how they have been raised above the sea. It is now obvious that different crustal blocks have moved toward each other, crushing and compressing strata caught between them; it is not so obvious why these blocks moved.

In recent years, scientists using sophisticated electronic devices have learned much about the physical properties of the earth, such as its magnetic, gravitational, and heat-flow properties. Much of this information has been gathered from the oceans. Sediments and rocks from the ocean floor can now be sampled by drilling, and the exact ages of many rocks can be determined by measuring their radioactive elements. These new data from the ocean floor and the details of rock structure on the land have been unified in the theory of *plate tectonics*.

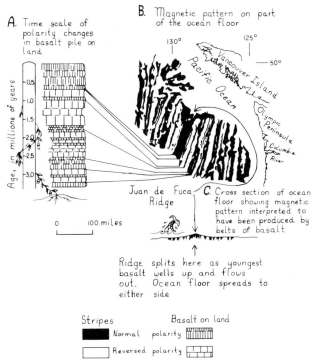

A. Time scale of polarity changes in basalt pile on land

B. Magnetic pattern on part of the ocean floor

Juan de Fuca Ridge

0   100 miles

C. Cross section of ocean floor showing magnetic pattern interpreted to have been produced by belts of basalt

↑
Ridge splits here as youngest basalt wells up and flows out. Ocean floor spreads to either side

Stripes   Basalt on land

■ Normal polarity

☐ Reversed polarity

Fig. 23. Comparing the magnetic pattern of basalts on the land to magnetic striping on the ocean bottom. After Cox, "Geomagnetic Reversals" (1969), and Raft and Mason, "Magnetic Survey off the West Coast of North America" (1961)

Rock Magnetics. Most rocks that are rich in iron, and basaltic rocks in particular, are naturally magnetic. Evidently, when a melted rock (such as a lava flow) cools, the atomic characteristics of its iron-bearing minerals are influenced by the earth's magnetic field; the rock becomes a very weak magnet, and its poles are aligned with the earth's field. In the geologic past the earth has periodically reversed polarity, that is, the north magnetic pole has switched places with the south pole. Lavas that erupted during a time of reversed polarity retain reversed polarity; those that erupted when the earth's polarity was like today's, which is called "normal," retain normal polarity. The north and south poles of the lavas can be detected with a sensitive instrument. Working in lava piles in areas of little crustal disturbance, geologists have been able to develop a schedule or time scale of these past polarity reversals (fig. 23A).

Ocean Floor Patterns. While geologists and geophysicists were climbing about the lava beds on land, oceanographers were cruising back and forth across the ocean, measuring the magnetic properties of rocks on the ocean floor. Although they could measure only the grossest kinds of magnetic properties from the surface of the ocean, their traverses revealed a startling striped pattern of magnetic changes. The pattern parallels prominent ridges on the ocean floor; furthermore, the pattern on one side of the ridges is a mirror image of the pattern on the other side (fig. 23B). The ridges had long been recognized as impressive, if little understood, features of all the world's oceans.

Amazingly, the pattern of magnetic changes on the sea floor match the vertical magnetic reversal patterns of progressively older lava layers on land. Thus the ocean floor stripes seem to represent the same polarity reversals found in the lavas on the land.

The Spreading Ocean Floor Meets the Continent. As geologists began to realize that the ocean floor was progressively older farther away from the ridges, they concluded that new crustal rock must be forming at the ridges and moving out to either side. Knowing that the oceanic crust is basalt, at least on the surface, scientists realized that molten rock had been added to the ridge as the ridge pulled apart at the center. Some of the rock erupted as undersea lava flows, but most cooled deep in the crack of the split ridge. It seems as if the crack in the oceanic crust, although continually opening, is also continually caulked by hot rock from below. The process has gone on so long that much of the oceanic crust is made up of long strips of caulking, and these strips take on the magnetic polarity of the earth's field, normal or reversed, as they form (fig. 23C).

But if the floor moves away from the ridges, where does it go? The striped pattern of magnetic variation ends rather abruptly against the continents and in some places appears to have been swept under them. The distribution of earthquakes deep beneath the margins of some continents suggests a possible answer. The points where these deep earthquakes originate crudely define planes dipping deep under the continents. These planes seem to represent the moving plates of oceanic crust that are moving down beneath the continental plates (fig. 24).

Using the plate tectonic scheme, we can visualize the earth's whole crust as consisting of gigantic plates, moving away from each other at the oceanic ridges and riding over or colliding with each other along continental margins. The area between the ridge and the continental margin is one plate, and the continent is another.

Deep-seated earthquakes that might define the boundary between the oceanic and continental plates off the coast of Washington have not been recorded. Nevertheless, the magnetic pattern suggests that much of the Pacific oceanic crust plunged beneath the North Amer-

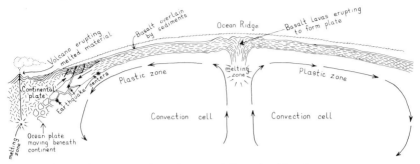

Fig. 24. Sketch of plate tectonics scheme and disruption of rocks

ican continent. Some geologists attribute this lack of earthquakes to a lull or cessation in spreading in the northeast Pacific at the present time.

Currents Deep in the Earth. The driving force for the movement of these gigantic crustal plates is not known, but an old theory of *convection currents* in the earth may fit into the new plate tectonic scheme. The interior of the earth can be likened to a heated pot of soup in which a convection current develops when the hot (that is, less dense) soup on the bottom rises to the top, forcing the cooler, heavier soup to the side of the pot, where it descends. If convection currents are moving within the earth, the relatively rigid and cool crustal plates of rock on the earth's surface may be moved to and fro. Some of these plates, like the masses making up the spreading ocean floor, will grow at the ridges and be consumed at the continents (fig. 24). Others, like some continents, may drift passively on the slowly moving substratum. The configuration of the earth's continents and ocean basins may change continually throughout geologic time.

Colliding Plates and Olympic Rocks

Although much is to be learned about the earth before the new theory of plate tectonics can be proven correct, it explains how rocks are deformed, and it can explain many features seen today, including many, though not all, of those we find in rocks of the Olympics.

Where two plates collide, such as at a continental margin, the rocks are likely to be squeezed and mashed to a remarkable degree. Olympic rocks offer abundant evidence of such folding, smashing, and tearing apart on a colossal scale (see fig. 25, and notes 9, 35, 105, 125, 172, among others). The disrupted rock belts of the Olympic core (fig. 16) are long and thin; even though they have been highly deformed, they maintain their continuity, as might be expected where a broad oceanic plate moves under a long section of the continent's

Fig. 25. Partly disrupted sandstone beds in slate on west side of Mount Fitzhenry. Mount Angeles in background and Mount Baker volcano in the far distance

edge. Also, a crude arrangement of younger and younger rocks westward suggests a sequential collision of material with the continent. The youngest rocks, farthest to the west, were not so strongly smashed as those farther east, possibly because the rate of motion between colliding ocean and continental plates was slowing down.

The basalt of the Olympic horseshoe was once riding on the ocean bottom. It may have erupted not far from the continent's edge and been skimmed off the oceanic plate as it plunged under the continent; it then stuck to the margin of the continent. Rocks that had been deposited either on top of the basalt pile or on its continental side (that is, the peripheral rocks; see fig. 26) were folded along with the basalt pile, although its great mass and rigidity protected them from extreme deformation. Not so for rocks deposited seaward of the pile, some of which were just the thin oceanward edge of the same basaltic pile. They were crammed against and underneath the main basalt abutment. This material was not only folded and sliced as it collided with the pile but was also pushed down to regions of higher temperatures, where new minerals began to form in it (note 136).

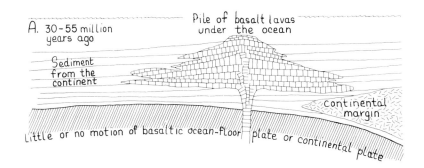

A. 30-55 million years ago

Pile of basalt lavas under the ocean

Sediment from the continent

Continental margin

Little or no motion of basaltic ocean-floor plate or continental plate

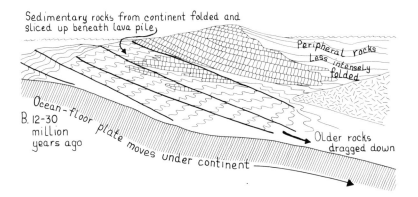

Sedimentary rocks from continent folded and sliced up beneath lava pile

Peripheral rocks less intensely folded

B. 12-30 million years ago

Ocean-floor plate moves under continent

Older rocks dragged down

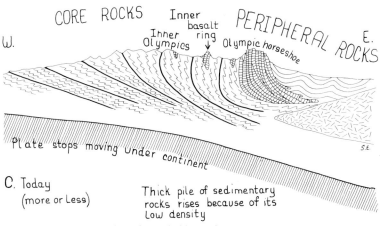

CORE ROCKS

Inner basalt
Inner ring
Inner Olympics
Olympic horseshoe

PERIPHERAL ROCKS

W.

E.

Plate stops moving under continent

S.E.

C. Today
(more or less)

Thick pile of sedimentary rocks rises because of its low density

Fig. 26. Development of the Olympic Mountains

We still do not know why the Crescent Formation and the arcuate belts of core rock are arranged in the horseshoe pattern. If we view the basaltic horseshoe from high above, we see that it appears to bend into an inside corner formed by the older geologic terranes of Vancouver Island and the Cascade Range. It looks as though it were squeezed into the corner (fig. 27), but there is little geologic evidence to support such an assumption.

However the horseshoe formed, the thick mass of mostly sedimentary rock from the continent was jammed up against and under the edge of the basalt, which in turn was jammed against the continent. When movement of the ocean floor ceased, the sedimentary rock began to rise because the sandstone and shales are lighter than the oceanic crust beneath. The spreading ocean crust acted as a conveyor belt, driving the rocks down beneath the continent; but when movement ceased, the rocks bobbed up like a cork (fig. 26C). This last episode of bobbing, accompanied by additional disruption of beds and faulting, raised the rocks in a domelike fashion to produce the height of land we call the Olympic Mountains. The mountains we see today were carved by erosion; even as the land rose, streams and rivers began to carry the rocks back to the sea.

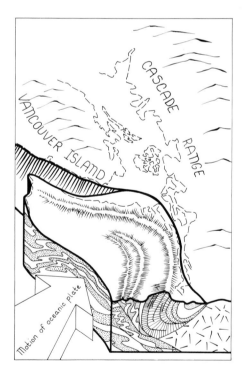

Fig. 27. Diagram showing how the rocks that form the Olympic Mountains may have been squeezed into the inside corner of Vancouver Island and the Cascade Range by the opposing movement of the oceanic and continental plates

Making Mountains
Out of Rocks

The story of the Olympic Mountains is long and complex. Although the material of which the mountains are made, the rock, is important in determining their nature and appearance, the forces that have sculpted the rock play the greatest role. The most interesting and varied rocks in the world uplifted as a high, uneroded plateau will not offer much to the mountain lover. It is the happy combination of rocks, uplift, and vigorous erosion that create the mountain scene.

Erosion

"How many years must a mountain exist
before it's washed to the sea?"
"Blowin' in the Wind" Bob Dylan*

Three main erosional agents have carved the Olympic Mountains: running water, glacial ice, and the direct action of gravity. They work in concert to carry the mountains back to the sea from which they were born.

Also of some consequence to the mountain scene is weathering, which prepares the rock for its erosional trip. *Chemical weathering* is particularly important in the southwestern Olympics, where hard basalt and softer sandstone and shale are all converted to clay by chemical action under the thick forest humus on low hills. Rocks so weathered are commonly stained red, red-brown, or yellow by iron oxides, which are chemically weathered from iron-bearing minerals. Roadcuts in the southwestern Olympics are characteristically colorful.

In the high country of the Olympics and in the lowlands of the north and east sides, clays produced by deep chemical weathering are not noticeable because the rocks there have been scraped clean by glaciers. Furthermore, in the high country, *mechanical weathering,* such as the fracturing of rocks by repeated freezing and thawing of water in cracks (note 73), is much more important than it is at lower elevations.

Running Water: The Persistent Sculptor. Anyone who has stood at the brink of an Olympic gorge and watched the foaming river at the bottom knows that running water is king in this land. From the very instant that the part of the earth's crust that is now the Olympics pushed up above the sea, rain and melted snow have been carrying bits of it back down again. As the land rose and the growing mountains intercepted more moisture, trickles grew to streams, and streams grew into rivers. The running water cut deeper and deeper into the rising land mass. Several factors contribute to a river's ability to cut: the amount of water it carries, the steepness of its descent, and, very important, the load of silt, sand, and gravel, which are its cutting tools. For example, an Olympic river that is laden with silt, sand, and gravel supplied by glaciers cuts bedrock faster than does a clear river not fed by glaciers.

The radial drainage pattern of the Olympics results from the domal uplift: as the Olympic land mass bowed up from the sea, water tended to run off in all directions. Rivers that had some advantage over their neighbors in rate of erosion extended headward more rapidly and captured more drainage area. The Elwha, which appears to breach the very center of the uplift (fig. 28), may have captured more drainage than the other rivers because it had a shorter, steeper course to the sea. The eastern rivers are short today, but they must have had a longer trip to the sea before the Cordilleran ice sheet carved out Hood Canal (note 1 and below).

With the major drainage pattern set, the erosional forces could begin shaping the mountains in a variety of interesting and subtle ways. Of utmost importance to erosion by running water are the differences in rock hardness. The basalt of the Olympics is generally harder than the surrounding shale and sandstone; it resists erosion and stands out in bolder relief, making sharper peaks and pinnacles (notes 17, 46). The sandstone is harder than the shale or slate, so it too becomes etched out. When interbedded with shale or slate, it commonly forms ribbed cliffs (notes 55, 135).

This *differential erosion* has modified the original radial drainage pattern. Because they cut faster in soft rocks than harder ones, some tributaries of the main radial streams developed deeper valleys in the softer rocks between hard basalt and sandstone layers than did the small streams crossing the basalt. Eventually the streams in the deeper valleys between the hard rocks captured the drainage of the smaller streams. Thus, the drainage pattern in some areas of the Olympics was modified from radial to radial and concentric (fig. 29).

In the core rocks, control of erosion by rock hardness is much more complex than around the periphery and of course has been consider-

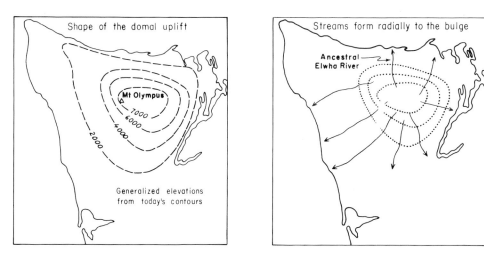

Shape of the domal uplift

Mt Olympus
7000
6000
4000
2000

Generalized elevations
from today's contours

Streams form radially to the bulge

Ancestral
Elwha River

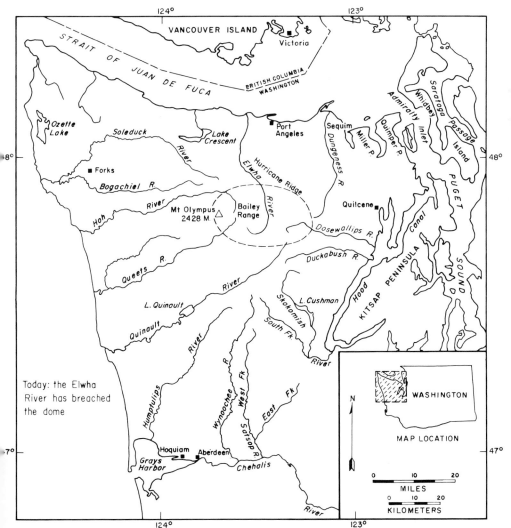

Today: the Elwha
River has breached
the dome

Fig. 28. Development of the basic drainage pattern of the Olympics

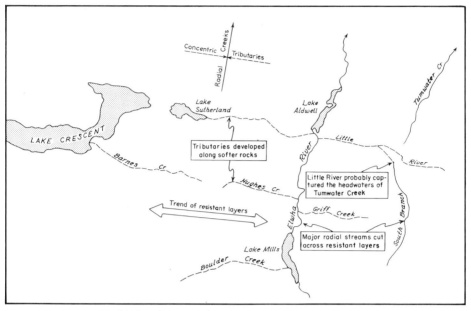

Fig. 29. Development of radial and concentric drainage

Fig. 30. Glacial scene reminiscent of the Ice Age. The southern Bailey Range and Mount Olympus massif, viewed from near Bear Pass

ably modified by glaciers. Rocks of the core have a strong northwest structural grain. Many factors contribute to this grain, but predominating are aligned beds of sandstone and basalt, mostly steeply tilted, and cleavage in slates (note 64). The drainage pattern clearly parallels the northwest grain (see relief map), but other structures in the rocks such as joints and faults are also influential.

Running water finds all these subtle differences and carves the softest zones, layers, and areas into rills, gullies, and valleys. The harder areas stand out as ridges and peaks.

Glaciers: The Heavier Hand. Although rivers established the grand pattern of Olympic scenery, and running water is the underlying melody of the Olympics today, the stillness of snow and glacial ice once filled the land, and glaciers of today are still an important part of the high mountain scene.

Whenever the accumulation of winter snow exceeds the amount that melts in summer and gets thick enough, it becomes ice. If the ice grows thick enough, it becomes plastic on the bottom under the weight of the snow and ice on top and begins to flow as a glacier. Many times in the geologic past glaciers grew to immense sizes. The most recent major icing of the earth occurred during the Pleistocene Epoch, beginning about 2 million years ago. During the great Ice Age, ice sheets spread from the higher latitudes into many areas inhabited by man today. At the same time smaller ice sheets and innumerable alpine glaciers formed in mountains all over the world. The Olympic Mountains not only bore their own system of extensive alpine glaciers but were nearly surrounded by the vast Cordilleran ice sheet that expanded out of western Canada (fig. 30).

Glaciers are important sculptors of the land, and they do their work in ways that differ from the ways of a river. They move very slowly but can carry great amounts of rock debris; individual blocks can be as big as houses (note 24, 164, 182). Debris carried by rivers is sorted by variations in the velocity of the water; mud, sand, and gravel are deposited separately. Glaciers do no sorting. When they melt away from their load of debris, it all falls together, forming an unsorted pile or *moraine* (notes 139, 159), unless water from the melting ice has a chance to sort it.

A river occupies only a small U-shaped channel in the bottom of the V-shaped valley it carves (fig. 31), but the glacier carves the whole valley into one great U-shaped channel (notes 80, 102, 131, 161). The head of a river valley generally merges imperceptibly with the ridges or highlands, whereas the head of a glacier valley is a steep-walled bowl, called a *glacial cirque* (notes 14, 44, 53). Rivers abrade with sand and silt, but glaciers scrape their channels with sand plus coarse rocks, held firmly and pressed down by the weight of the overlying ice. Glacier-carved rock can be as smooth as a polished headstone or cut by deep grooves and gouges, or both (note 53 and fig. 9). A river generally tends to be more sensitive to slight variations in rock hardness than the less selective glacier. Glaciers, like bulldozers, are not strongly influenced by the subtle differences in nature's makeup.

Early Olympic geologists were so involved with solid rocks and somewhat obscure events of the distant geologic past that they gave little time to the more obvious records of earth history written in the loose deposits of the Ice Age. In 1913, J Harlen Bretz, who might be called the grand old man of northwestern Pleistocene geology, described the Olympic Mountains as they were during the Ice Age, surrounded on three sides by a river of ice from Canada (notes 1, 56). Since Bretz wrote, many people—University of Washington students

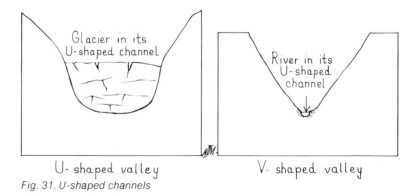

Fig. 31. U-shaped channels

in particular—have worked out the complicated story of the lowland glaciation.

At least six times during the Ice Age, the Puget lobe of the Cordilleran ice sheet crept down into the Puget Sound area. The ice dammed up against the Olympics and split into two branches: one branch flowed out the Strait of Juan de Fuca to the sea, and the other flowed down the east side and part way around the southern end of the mountains. The ice was at least 3,500 feet thick on the northern edge of the mountains, where it smoothed down the foothills and gouged out the peripheral valleys (notes 56, 112, 116). Furthermore, the Cordilleran ice, loaded with rock debris picked up in Canada, dropped foreign rocks or *erratics,* such as granite, gneiss, and schist, all around the northern, eastern, and southern sides of the range.

In the earlier part of the Ice Age, the Olympic massif may have been almost covered by its own cap of ice. That part of the geologic record is unclear. It is known that while the Cordilleran ice sheet was stretching southward, snow piling deep on the high peaks of the Olympics grew into glaciers that descended most Olympic valleys at least four times. These glaciers gouged out the stream valleys, and melt water deposited tremendous amounts of outwash gravels around the west and south sides of the mountains (notes 128, 142).

The alpine glaciers on the northeastern side of the range melted back before the last retreat of the Cordilleran ice. The valleys, thus dammed by the ice sheet, were filled with fiordlike lakes. Icebergs that broke off the ice sheet floated into the interior of the mountains. When these bergs melted, they dropped erratics far up the drowned valleys (notes 56, 96).

Deciphering the complex history of alpine glacial advance and retreat employs many scientific disciplines. Geologists such as Dwight Crandell of the United States Geological Survey have sorted out moraines and glacial outwash on the Hoh, Queets, and Quinault rivers. Old moraines can be told from younger ones because their boulders and cobbles are more deeply weathered. The ages of young moraines may also be determined by counting the annual growth rings of the trees on them. Experts examine fossil pollens in layers of mud and silt that were deposited in ancient bogs. Professor Calvin Heusser of New York University has been able to estimate the abundance of different plants present on the western Olympic Peninsula at various times of glacial advance and retreat. By comparing these abundances with those of present-day plant communities at sites ranging from the Olympics to the Arctic, he has been able to describe the changes from tundra to forest and back again as the glaciers waxed and waned and has estimated the average July temperatures, which have been only three to eleven degrees (Fahrenheit) colder than today. Students from the University of Washington have detailed the advance and retreat of the glaciers of the Queets, Quinault, Humptulips, and Wynoochee, as well as other drainages.

Olympic alpine glaciers are not only noted for their beauty; they have also contributed much to man's understanding of his environment. Glaciologists have been watching changes in the Blue Glacier on Mount Olympus and its relationship to climate for many years. Professors Robert P. Sharp, Barkley Kamb, and their associates have measured the pulse, form, and flow of the lower Blue Glacier (note 134). Professor Edward LaChapelle and his group have dissected the Snow Dome on the upper Blue. The research teams have determined the ways in which snow accumulates and ice melts, how the glacier flows, and even variations in the types of oxygen (isotopes) making up the ice. For instance, the Blue Glacier moves by internal flow—that is, slippage on planes in the ice crystals themselves—and by slippage along the bottom. There is more flow in the center of the glacier than at the bottom and along the edges. Differences in flow velocity stretch the brittle ice near the surface and it cracks, forming crevasses.

The world became warmer than it is today after the last ice advance in the Puget Sound area about thirteen thousand years ago. Some geologists say that this warm period was a minor fluctuation and that we have not yet emerged from the Ice Age because the earth is still somewhat iced, as shown by the Antarctic and Greenland ice caps, the frozen North Polar Sea, the numerous small glaciers sprinkled about many mountain ranges, and the wide expanses of permanently frozen ground in Canada, Alaska, and Siberia. During the warm period the northern ice sheets and most of the alpine glaciers disappeared, although the largest and highest—such as the Blue and Hoh glaciers —may have survived. Present-day Olympic glaciers apparently were born about twenty-five hundred years ago and reached their maximum growth only about two hundred years ago, evidence of a current cool episode. Since then they have been steadily retreating, but whether this present warming is a long-time trend or whether a full-scale ice age will recur is unknown.

Gravity: The Great Leveler. In the long run, gravity is the force behind all the processes that tend to drag the mountains back to the sea. The spherical earth is a restless creature whose internal troubles keep wrinkling up its skin, but because it is so massive, that thing we call gravity keeps smoothing out the wrinkles, trying to reform the earth into a perfect sphere.

In the Olympics, gravity often moves soil and rock downhill with only minor help from water or ice. Slow movement is called *creep* and fast movement, *landslide* and *rockfall*. Creep is commonly shallow: the surficial cover of weathered rock, moraine, or soil slides slowly downhill, wrinkling like a rug on a tilted floor (notes 68, 71). Deep creep—whereby masses of bedrock sag downhill on steep mountain-sides—has produced ridgetop depressions, cracks, and hillside swales throughout much of the high country (notes 11, 68, 109, 121). Landslide material, on the other hand, usually breaks away in a great rush, sliding or flowing into the valley below (notes 13, 15, 89, 112, 140). Rockfalls build slopes of coarse debris called *talus* at the bases of cliffs.

The soft rocks of the Olympics appear to have crept and slid valley-ward most actively right after glacial retreat, when the walls of the deeply carved valley channels were most unstable (notes 11, 140). But even today a creek or river gnawing into the toe of a slope may remove support from upslope material, causing the whole mass to slide gradually or catastrophically down (note 89).

Creep and landslide processes tend to smooth steep and jagged ridges into low, rounded hills. They turn the steep-walled gorges sawed by rivers into V-shaped valleys (note 176) and destroy the U shape of glacier channels. The smoother and lower the hills get, the slower the descent of rock and debris to the rivers and the slower the removal of the material to the sea by the rivers. But gravity never relents and, given enough time, will smooth even the boldest peaks of the Olympics into low, rolling hills.

Rising Land and the Olympic Scene

Fortunately for Olympic visitors, the restless crust of the earth has not yet allowed gravity and its cohorts, water and ice, to level the land. There is considerable evidence that the Olympics today are less gentle than they once were. As indicated earlier, when a mass of the earth's crust is uplifted high above the ocean, streams are born and begin gnawing away at the elevated land. They work rapidly, creating steep, V-shaped valleys and high, jagged ridge crests. Eventually the valleys reach a maximum depth, and the crests a maximum sharp-ness; the streams cannot cut deeper than the level of the sea to which they flow. Then the slower erosive processes such as creep, land-sliding, and the washing of rain round off the sharp ridges and flatten the steep valley sides.

At some time in the past, the ridges of the Olympics were smoothed into rolling meadowlands, such as are now seen on Hurricane Ridge (note 70) and on a few ridges along the eastern side of the park. But before the agents of rounding and smoothing had a chance to com-pletely reduce the mountains to lowland hills, the range was glaciated and uplifted again. In general, glaciers are even more ravenous than young streams in the way they eat at mountainsides, leaving their steep-sided cirques. They have also been effective in removing all traces of the old, gentle upland in the central part of the range and on the north sides of ridges. The uplift, however, gave the streams new appetite by increasing their gradients and they once again began steepening the valleys and sharpening the ridges. The gorges along many of the major rivers are testimony to the steepened gradient. The river erosion has been so rapid that slower gravitational processes, which would otherwise open the gorges into V shapes, have not yet caught up. The story might be further complicated had sea level changed: lowering the sea level has the same effect on streams as raising the land. But most geologists feel that the oceans are as full today as they have been for many years.

Considerable uplift of Olympic rocks results from their tendency to float on the deeper and denser oceanic crust; but much recent uplift may have been caused by the removal of the Cordilleran ice sheet, which once depressed the land by its great weight. The high outwash terraces of the south and west sides of the mountains may be evidence of a time when the land was depressed and streams running slowly to the ocean could not carry all the debris delivered to them by the melting glaciers. The ice sheet does not provide the complete answer, however, for what goes down must have been up; the land must have also been higher *before* the coming of the ice sheet. In fact, when we consider all the advances and retreats of the ice sheet, it seems that the land must have risen and fallen numerous times.

The sum total of the movements, however, seems to be up, since the mountains are still high and the streams are aggressively eroding them. The mountains may still be rising, but whether they can gain elevation over the effect of erosive processes will be known only by surveyors of the future, provided that man continues to be concerned with such things.

PART II
Notes on the Geology
of Olympic National Park

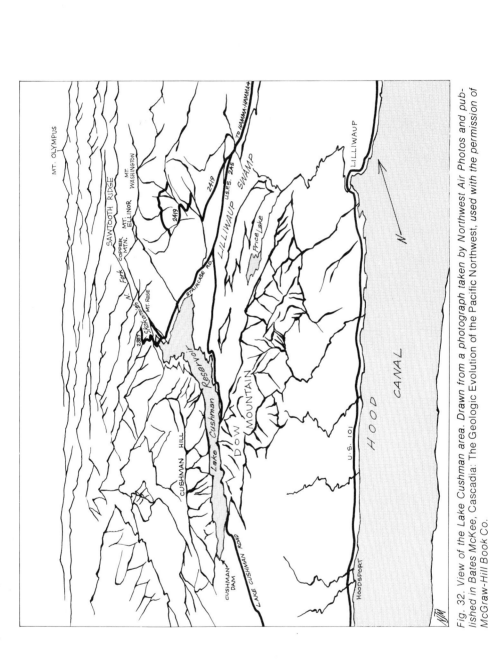

Fig. 32. View of the Lake Cushman area. Drawn from a photograph taken by Northwest Air Photos and published in Bates McKee, Cascadia: The Geologic Evolution of the Pacific Northwest, used with the permission of McGraw-Hill Book Co.

Eastern Approaches

Lake Cushman–North Fork of the Skokomish River

Lake Cushman is a hydroelectric reservoir that fills a glaciated valley. (1)
Before the dam was built, a smaller Lake Cushman was dammed by
moraine deposited at the margin of the Cordilleran ice sheet, which
pushed up against the base of Mount Ellinor and Mount Washington.
The older lake filled only the northwest end of the present lake (fig.
32). Pioneer geologist J Harlen Bretz wrote in 1913: "From the
so-called 'Half Way Rock' along the [Mount Ellinor] trail, it requires
only a slight effort of the imagination to see below one, the
dammed-back Skokomish Glacier opposed by the mightier Puget
Sound ice mass which crowded westward against the mountains."

The moraines that Bretz examined are covered by Lake Cushman
reservoir, but we can still view this scene of ancient glaciation from
the Big Creek Road (F.S. 2419) or, like Bretz, from the upper parts of
the Mount Ellinor Trail, which leaves the Big Creek Road 5.0 miles up
from F.S. 245. Between Dow Mountain (fig. 32) and the main range,
the flat, tree-covered land is covered with gravels left by the ice cap.

About a mile up a logging road (F.S. 2357) that crosses the upper (2)
end of Lake Cushman on a causeway are some spectacular cuts of
dark-green pillow basalt containing garish red beds of limestone.
Robert Garrison has shown that similar beds are partly composed of
millions of submicroscopic skeletons of plankton (called coccoliths),
one-celled animals that floated on the ocean some fifty to seventy million years ago (fig. 33). These fossils and globular basalt pillows (also
seen along Lake Cushman and elsewhere—see notes 25, 46, 60)
prove that the basalt flowed out on the floor of the ocean.

At a spot 1.3 miles from the main North Fork Road a dike of diabase
intruded as molten rock into a crack in the basalt and red limestone
(fig. 34). The diabase is darker green than the basalt and looks grainy
on the surface. On freshly broken pieces small crystal faces glitter in
the sun (fig. 49 and note 46). The margin is dense or glassy looking
compared with the grainy material of the center. The margin cooled so

Fig. 33. Electron micrograph image of plankton skeleton from red limestone, Olympic National Park, enlarged about 6,450 times. Courtesy of R. Garrison

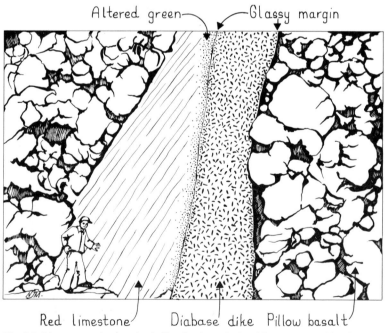

Fig. 34. Diagrammatic sketch of dike that intruded red limestone and basalt

quickly as the molten rock came into contact with the cooler basalt and limestone that there was no time for large crystals to grow.

Where the dike cuts the red limestone, the heat changed the red rock to green. The limestone was initially red because it contained tiny scattered crystals of hematite (iron oxide). During baking by the dike, the hematite combined with the limestone (calcium carbonate), mud and sand (silica), and water in the pore spaces of the rock to become epidote, a green mineral made of calcium, iron, aluminum, silica, and water.

Views from the higher parts of F.S. 2357 are well worth the haz- (3) ardous drive past logging trucks. The whole upper North Fork of the Skokomish can be seen, its retinue of steep, forested ridges and snow-patched peaks beyond, as well as the rocky basalt points of Mount Lincoln, Copper Mountain, and Mount Ellinor.

On the southwest side of the North Fork of the Skokomish, just (4) above the fine pool by the Staircase Campground, is an old tunnel dug by prospectors in search of manganese and copper. The Shady Lane Trail crosses this old digging on a bridge. This excavation and others nearby are in red limestone and basalt. In many places in the Olympics, manganese ores are found in close contact with the red limestone (see Chap. 1, "Basalt and Its Associated Rocks," and note 60); but here the prospectors were no doubt disappointed, for there is little evidence of large deposits.

On the steep hillside just above the conspicuous alluvial fan (note (5) 169), the Wagon Wheel Lake Trail passes an old prospect. Tiny spots of chalcopyrite (copper-iron sulfide) and pyrite in a diabase dike (note 2) and in the altered slate nearby seem to be what inspired some prospector to this great labor.

The beautiful, water-carved rocks of the cascades at Dolly Pool on (6) the Four Stream Trail (0.5 miles from Staircase) are a tough diabase mass (note 46) in softer slates and red limestones.

Dead Horse Hill is a moraine—assorted rock debris that the Sko- (7) komish Glacier left behind on its trip down the North Fork (see note 159).

The Sawtooth Range is one of the most rugged, rocky ridges of the (8) Olympics. Its blocks and spires of pillow basalt and breccia are fa- vorite challenges for the rock climber. The originally horizontal lava beds here stand on end, and the myriad of sharp summits—the saw's teeth—are formed by concentrated erosion along parallel cracks, or *joints,* that cut across the vertical beds of basalt (fig. 35).

Gladys Divide is a rubble-filled notch cut in the highly disrupted (9) slate of a fault zone stretching northeastward across the upper Hamma Hamma and beyond to Hagen Lake (see note 18 and fig. 17).

Fig. 35. Beds and joints form pinnacles on Sawtooth Ridge. View is south from Gladys Divide Trail

The sandstones and slates on Mount Gladys are also highly disrupted, and the hiker traversing the rocky ribs and verdant nooks of
the mountain will find contorted, pulled-apart sandstone beds of every
shape and size (see also note 35).

(10)   Over sixty years ago, prospectors and miners were exploring the
upper North Fork for valuable minerals. A claim was staked in 1907
near Black and White Lakes, on outcrops bearing native copper and
minerals of copper and manganese. Manganese minerals were mined
about 1918, and 100 tons of ore were carried out by mule train. The
diggings are easily seen above the trail as it comes up to the lakes
from below. An early Olympic geologist, Ralph Arnold, described
these ore deposits and, in a footnote, the names of the lakes: "Mr.
Stanard furnishes the following note concerning the origin of 'Camp
Black and White': This camp was named by some of the early elk hun-
ters from a brand of whiskey of that name, one of the party being
sober enough at one period of their sojourn at this place to mark the
name prominently on a tree."

Perhaps no place in the Olympics more forcefully impresses the (11)
hiker with the phenomenon of ridgetop depressions than does the Six
Ridge Trail. After gaining the ridge crest, the traveler climbs up and
then down, through and around sharp "valleylets" and defiles that
follow the ridgetop or cross it at a small angle. In some of the depressions, snow in the undrained bottoms lingers long into August. When
the valley glaciers of Six and Seven streams melted back long ago,
they no longer supported the oversteepened valley walls, which
began to collapse. The depressions formed behind masses of rock
creeping valleyward (notes 121, 140).

The pass between Mount Hopper and Mount Steel is carved from (12)
disrupted slate and scattered blocks of broken sandstone. The lowest
pass, west of the trail crossing, leads to a flat-floored, meadowed
basin above the Duckabush that is sprinkled with large blocks of
greenish, metamorphosed basalt (greenstone). The basalt has fallen
from the cliffs of Mount Steel (to the west), where the keen-eyed traveler can see large masses of hard whitish sandstone and greenstone
smeared into the slates (see notes 35, 172).

Hamma Hamma River

Jefferson Lake formed when a huge landslide created a natural (13)
dam. Look for blocks of basalt on the hummocky surface of this landslide where the Jefferson Creek Road (F.S. 2401) climbs up to the
west side of the lake (6.3 miles from the main south side Hamma
Hamma River Road, F.S. 248). A less obvious slide also dams Elk
Lake. Such slides were probably common after the valley glaciers of
the Ice Age retreated from their channels (note 140).

Take the righthand fork of the upper Jefferson Creek Road (F.S. (14)
2401) for spectacular views (9.8 miles from F.S. 248) of the glacial
cirque between Mount Pershing and Mount Washington.

Lena Lake occupies a depression dammed by a huge landslide. (15)
The trail to the lake crosses Lena Creek amid the huge blocks of this
slide, and the southern end of the lake is littered with more monoliths
(note 13).

Upper Lena Lake was carved by a glacier from highly disrupted (16)
slate and sandstone, well displayed on peninsulas and points around
the lake. To the south, Mount Bretherton and associated craggy basalt
summits rise abruptly from the sedimentary rocks of the basin (fig.
36).

The logging road up Boulder Creek (a spur road off of F.S. 249) (17)
provides some views of the eastern Olympic peaks seldom seen from
roads. Mount Pershing and Mount Cruiser are rugged, resistant
masses of basalt left high by erosion, which has carved down the less

Fig. 36. Upper Lena Lake carved from sandstone and slate. Behind the lake
rise the basalt summits of Mount Bretherton

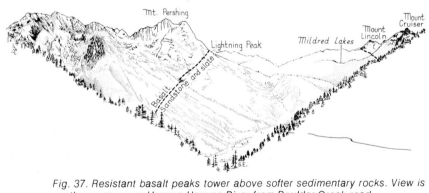

Fig. 37. Resistant basalt peaks tower above softer sedimentary rocks. View is
south across upper Hamma Hamma River from Boulder Creek road

resistant sandstone and slates lying between them (fig. 37).

The Putvin Way Trail up Whitehorse Creek (which leaves the (18)
Boulder Creek road 1.3 miles from the main Hamma Hamma Road)
leads to flowered meadows and waterfalls descending the stepped
cirques between Mount Skokomish and Mount Stone. A climb to the
saddle on the ridge between Stone and Skokomish gives a grand view
of Olympic geology. To the east are the basaltic peaks of Mount
Bretherton, Mount Jefferson ridge, Mount Pershing, and Mount Washington. These rugged peaks are part of the Olympic basaltic horseshoe (fig. 17). The rocks of Whitehorse Creek basin are sandstones of
the inner Olympics. The thick beds are standing vertically and even
lean over, upside down, to the east. Mount Stone and the ridge to the
south of the pass are parts of an inner basalt ring. West of this basalt
is a highly disrupted slaty zone, actually a wide fault zone, that has
been eroded out (note 32) to make a long swale west of Mount Stone
and extending southward across a small saddle below Mount Skokomish and beyond (fig. 38 and note 9). Across this swale are prominent exposures of intensely contorted black slate, studded with water-
melon-sized pieces of sandstone. This fault zone, an important part
of the southern fault belt (fig. 17), separates the disrupted sandstones
and slates seen to the west in Mounts Steel, Duckabush, Hopper, and

*Fig. 38. Erosion of disrupted slate zone produced this swale at the head of the
Skokomish North Fork. Mount Stone is on right, the two peaks of Mount Constance, center skyline, and Mount La Crosse, on left*

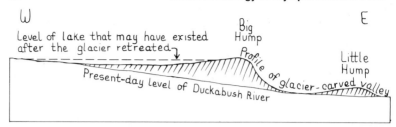

Fig. 39. Profile of Duckabush River near Big Hump

Anderson from the less deformed inner basalt ring and the outer basaltic horseshoe.

(19)   Mildred Lakes are tarns, or glacier-carved basins, set in Hudsonian
forest (see fig. 37). The lake basins have been scraped out of
thick-bedded sandstones, well exposed where the primitive trail
climbs over the ridge to reach the basin. Around and in the lakes are
huge blocks of pillow basalt, erratics brought down from Sawtooth
Ridge by the glacier that gouged out the lake basins.

Duckabush River

(20)   Both Big and Little Hump on the Duckabush Trail are the risers of
glacial steps in the canyon, places where the Duckabush Glacier cut
very deeply (see notes 30, 133). The trail is forced to leave the river
bottom at Big Hump because after the glacier melted, the river cut a
deep gorge in the lower part of the step (fig. 39). Outcrops of glacier-smoothed basalt are conspicuous on the downstream side of Big
Hump where the trail approaches the top.

(21)   Look for pillow structures in basalt (note 25) where the trail hugs
cliffs and climbs from what looks to be the postglacial gorge of the
Duckabush (note 20) to the upper glaciated valley. This valley bottom,
at 1,300 feet, is lower than the summit of Big Hump (1,700). Shortly
after glacial retreat, the upper Duckabush Valley may have contained
a lake dammed by the rock sill of Big Hump (fig. 39). The river cut
through the sill at its south end and drained the lake.

(22)   Marmot Lake, Hart Lake, Lake La Crosse, and other little ponds
occupy glacier-carved depressions in small cirques. The glaciers that
once occupied this alpine region were tributaries to the Duckabush,
and with a little imagination, the hiker can visualize the ice fall that
once stretched from the lips of the lake basins to the larger glacier in
the valley below.

Dosewallips River

(23)   An abandoned spur road just east of the National Park boundary
(13.7 miles from U.S. 101) leads to the old Elkhorn manganese pros-

pects east of Bull Elk Creek. The road ends in brush beneath a pre-cipitous hillside where the prospectors scrambled around the cliffs to find small ledges of manganese minerals.

The first impressive cliff of basalt to be seen along the east side of (24) the Lake Constance Trail (about 1 mile from the road) is not a part of the mountain, as it first appears, but a building-sized block of basalt either fallen from cliffs above or dropped by the glacier that once descended the valley.

Lake Constance is surrounded by pillow basalt that erupted on the (25) ocean floor. The lava beds have been bent up to stand vertically, and erosion has cut into the thick pile. The climb to Crystal Pass north of the lake brings the hiker through one of the most spectacular displays of submarine lava in the Olympics. Look closely to see how the curi-ous globs and tubes branch from one another. Rock is an excellent insulator; when a glob of melted rock oozes into water, its outer sur-face cools quickly to a glassy rind that protects the hot interior and allows it to cool slowly. Sometimes the rind cracks and the lava, oozing out, buds forth as a new pillow. Other times long, tubelike structures are formed as the lava makes its own insulated conduit under the water (fig. 40).

Fig. 40. Tubes and pillows in nearly vertical basalt beds above Lake Constance

(26)   The road to Dosewallips Campground passes beneath a spectacular cliff of pillow basalt. The rounded forms of the pillows look like
the surface of a giant's cobblestone road, here upended.

The pile of basalt, the Crescent Formation, exposed along the road
from Hood Canal to near Dosewallips Campground is the thickest part
of the Olympic basaltic horseshoe (fig. 16). We do not know how
much the pile has been thickened by faults or folds because they are
difficult to find in the basalts; but if the beds were turned back to near
horizontal as they must have been deposited, the pile would be at
least ten or twelve miles high. Judging from modern-day oceanic volcanoes like Mauna Loa in Hawaii, the original pile may have been
only six miles or so thick and has thus been thickened by folding or
faulting. Because much of the rock was deposited under the ocean,
the volcano was not necessarily a huge mountain as seen from the
land. Furthermore, the weight of the pile probably depressed the
ocean floor as the pile grew.

(27)   The bridge over the West Fork of the Dosewallips River crosses a
spectacular slot cut in thick beds of sandstone. These same sandstone beds can be seen along the main fork trail where it climbs up to
the Constance Pass Trail junction. The beds, now standing on edge,
represent floods of sand swept onto the deep ocean floor in a dense
current of sand and water.

(28)   About 1 mile downstream from Diamond Meadow, the West Fork
Trail climbs over a broad, convex hump, notable for its open, piney
woods, its dryness, and its even declivity toward the river. The feature
is a large landslide, coated with alluvial fan material. The material is
more porous than the surrounding hillsides and young enough not to
have developed a heavy, impervious soil; water runs through it,
leaving it relatively dry and providing favorable growth conditions for
the pines.

(29)   Diamond Meadow is periodically buried under snow avalanches
from the brushy gully to the north, and the deep snow hinders tree
growth. Cross the West Fork to see boggy elk wallows at the mouth of
Elk Lick Creek presumably near the site of a hot spring reported ex-
tant here in the late 1920s. This spring may lie near a branch of the
fault zone bearing the better known Olympic Hot Springs (notes 82,
83, and fig. 16). To judge by its popularity with elk seeking salts, Elk
Lick Spring is probably still active, seeping out from beneath slumped
river bank material and forest debris.

(30)   Honeymoon Meadows is a small, lushly green clearing on a glacial
step. Below the meadows, a steep drop in the valley floor leads to the
gentler grade of the main valley. During the Ice Age, glaciers came
into the area of the meadows from all sides to join as a long ice river

descending the West Fork. Valley glaciers typically erode in steplike fashion, the treads and risers developing in response to changes in the amount of ice, such as below a tributary glacier junction, or differences in the rock's resistance to glacial excavation. Sandstone boulders, dumped over the edge of the step by the glacier, fill the river where it cascades down to the lower valley.

The black cliff above Honeymoon Meadows magnificently displays disrupted sandstone blocks in slate (note 35).

From Del Monte Ridge look north out over ridges of sandstone and (31) slate. The hiker stands here in the middle of a great rock fold which can be recognized only after careful study of the whole area (fig. 41). The geologist notes the position of bedding in the thick sandstones at various places and if he pays close attention to which side of the bed was up when it was deposited (notes 52, 123), he can piece together the pattern which reveals this giant twist in the earth's crust.

The slot between the basalt masses of Little Mystery and Mount (32) Mystery is known as the Gunsight. It is the erosional expression of a fault, that is, a crack in the earth's crust where a mass of rock on one side has moved relative to rock on the other side. Because the hard basalts here have been ground up by the movement and slivers of

Fig. 41. Simplified geologic map of Del Monte Ridge area. Drawn after Cady et al., Geology of the Tyler Peak Quadrangle, 1972

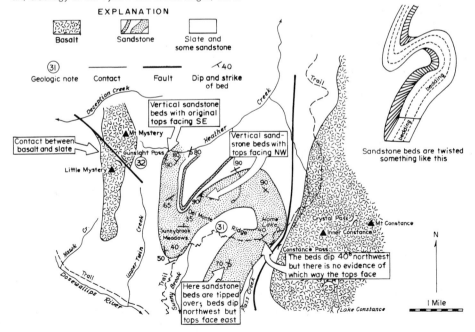

weak slate have been dragged in between the blocks of basalt, erosion has cut deeper in the fault zone than in the stronger rocks to either side. The small pinnacle of basalt in the center of the Gunsight, which looks like a sighting pin, is a sliver of hard basalt in the fault zone.

(33)   Where the trail climbs a broad, sloping shelf across from the mouth of Silt Creek, the Dosewallips is cut into a deep rock gorge. When the alpine glaciers came down the Dosewallips, before about 13,000 years ago, the main ice stream fed by the ice fields of Mount Anderson came down Silt Creek and cut down more efficiently than the smaller tributary that is now part of the Dosewallips above Silt Creek. The present upper Dosewallips River was once in a hanging valley (note 161), but since the glaciers retreated, the river has cut away almost all of the upper valley floor. Parts of the old floor remain just where the trail leaves the broad sloping shelf for the steep canyon side.

(34)   On the Dosewallips Trail, note the change in vegetation after crossing Deception Creek. The gravelly slope is well drained and supports only small trees of a pine forest rather than the denser growth of the preceding slaty canyons. The north bank of Deception Creek and the terrain to the north, almost to old Camp Marion, are of landslide material. The debris evidently came from greatly fractured sandstone beds high above Cub Creek.

(35)   West of the trail at Gray Wolf Pass, a black buttress of rock reveals some features testifying to the forces that have helped mold Olympic rocks. The cliffs are a mixture of slabs and angular pieces of thin sandstone beds in a mortar of slate. Look carefully to find pieces of folded sandstone beds. Figure 42 illustrates how such disrupted rocks formed (see also Chap. 1, "Colliding Plates and Olympic Rocks"). Similar zones of intense movement are widespread in the Olympic core, but the size of the broken pieces varies from small splinters most easily viewed with a microscope to whole mountains (notes 105, 125, 137).

Just south of Gray Wolf Pass is a small, ditchlike trough related to ridgetop depressions (fig. 43). This sidehill trough probably formed as the mountainside below it slid valleyward, as shown by bumpy, steplike topography below (see notes 121, 140).

(36)   At Dose Meadow the hiker ascending the Dosewallips Trail passes from terrane underlain by much sandstone and minor slate to a terrane underlain predominantly by disrupted slate. The transition is not obvious, but looking to the north above Dose Meadow, the observer can see cliffy ribs of sandstone on the east that end abruptly in steep but less cliffy slopes of slate below and to the west (see fig. 16).

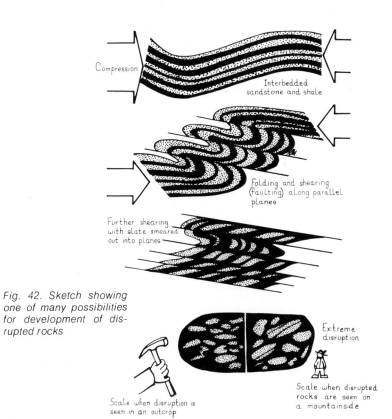

Fig. 42. Sketch showing one of many possibilities for development of disrupted rocks

Compression

Interbedded sandstone and shale

Folding and shearing (faulting) along parallel planes

Further shearing with slate smeared out into planes

Extreme disruption

Scale when disruption is seen in an outcrop

Scale when disrupted rocks are seen on a mountainside

Fig. 43. Slump depression near Gray Wolf Pass

The trail just west of Hayden Pass wanders along a shallow swale   (37)
on the ridgetop. Ridgetop depressions such as this are common in the
Olympics and in many alpine ranges around the world (notes 11,
121).

A small basin at the eastern base of Mount Claywood, easily   (38)
reached by climbing from the Hayden Pass Trail (along the east side
of the creek draining Claywood Lake), is filled with large blocks of
greenish basalt and small pieces of red limestone. The basalt forms a
prominent rib extending from Mount Fromme to the east shoulder of
Mount Claywood. A glacier hollowed out the basin of Claywood Lake,
and its moraine partly dams the lake. The red limestone is interbedded with gray sandstone in cliffs below the lake. The limestone
is rich in the microscopic skeletons of Foraminifera, one-celled animals that lived in the ocean. This is one of the few places in the core
of the Olympic Mountains where fossils are preserved well enough to
be identified, and they tell us that the rocks here are between 40 and
60 million years old.

The steep, airy trail up to Lost Pass traverses beds of sandstone   (39)
(note 36), and at Lost Peak the hiker sees a particularly thick mass of
sandstone beds (fig. 44). Because the sandstone beds are thick and
have only thin slate beds between them, they are more resistant to
erosion than the surrounding thin-bedded sandstone and slate and
thus stand up boldly in Lost Peak. Similar thick beds can be seen in
the peaks of Mount Cameron to the north.

At the base of the talus fans around the north side of Lost Peak is   (40)
a hummocky carpet of moraine, ending abruptly in a sinuous, partly
tree-clad ridge, a classic example of a terminal moraine (fig. 44). At
the lower, western end, concentric arcs in the moraine blanket show
successive stands of a glacier, now long gone (note 159). This small
glacier was but a puny child of winter snow, sheltered by the tall peak
from the sun in the southern sky. During a much earlier, colder time
(namely, the Pleistocene), the entire cirque at the head of Lost River
was filled with ice.

*Fig. 44. Basin at head of Lost River with Lost Peak on right, moraine ridge
center and right. View is east down the Dosewallips River: from left to right in
background are Martin Peak, Mounts Deception, Mystery, and Little Mystery of
the inner basalt ring; Mount Constance, skyline, right of center, of the basaltic
horseshoe. Aerial photo by Austin Post*

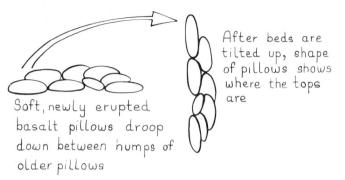

After beds are
tilted up, shape
of pillows shows
where the tops
are

Soft, newly erupted
basalt pillows droop
down between humps of
older pillows

Fig. 45. A way to tell which side of a tilted bed of basalt pillows was
originally up

Fig. 46. In Royal Basin, Royal Creek falls over a step of basalt

Northern Approaches

Dungeness River

Where the Dungeness River road swings in and out of gullies along   (41)
the precipitous southeast side of Maynard Peak, the traveler passes
through the bottom flows of an immense pile of basalt (the Crescent
Formation), deposited on the ocean floor between 40 and 55 million
years ago. The flows are tilted on edge here so that a trip up the river
is a trip back in time. Between Maynard Peak and Tyler Peak is a thick
sequence of shale and sandstone, but more basalt makes cliffs on the
southeast side of Tyler Peak, high above the main road. These basalts
are some of the earliest flows of the ancient volcanic outpouring.

Many of the exposures along the road display the pillows formed
when the lava erupted into the sea (notes 25, 26). Because the pillows
were hot and soft when formed, they tended to droop into depressions
in whatever lay beneath them. Commonly they filled depressions between the cold, humped backs of earlier-formed pillows, providing a
means for the geologist to tell the original top of the lava bed (fig. 45).

The falls of Royal Creek are held up partly by a "dam" of basalt   (42)
(fig. 46). The upended flow that extends from Gray Wolf Ridge on the
west to Royal Basin on the east resisted the scraping of the Royal
Creek Glacier and still resists the cutting of Royal Creek.

Several large creeks in the upper Royal Creek area issue from un-   (43)
derground channels. They are simply normal surface streams that flow
more easily under large permeable accumulations of basalt blocks
such as in talus or landslides.

The cirque at the head of Royal Creek is a delightful area of gla-   (44)
cier-carved basins and polished knobs (fig. 47). Ancient terminal
moraines (notes 139, 159) are conspicuous in many places, especially below the small dying glacier at the east face of Mount Deception. Royal Lake is a tarn, that is, a lake occupying a glacier-carved
depression. There may well have been more lakes in the basin at one
time, but they have turned into meadows as they filled with sediment
(note 120).

Fig. 47. Royal Basin. Mount Deception to the left, the defile of Surprise Pass and Mount Clark partly hidden behind the ridge in the middle ground and to the right, Royal Lake in foreground

Fig. 48. Well-developed joints in basalt and diabase on the south side of Mount Deception

Surprise Pass, the prominent notch through The Needles, has been (45)
eroded out of pulverized basalt along a fault. Movement of the blocks
of rock on either side of the pass have ground up the basalt in the
fault zone (note 32).

The high, rugged ridge extending from Little Mystery to The Nee- (46)
dles is a major segment of the inner basalt ring (figs. 16, 17). The
lavas that form much of this ridge oozed out on the ocean floor somewhat later than the lavas making up the main basaltic horseshoe
(Chap. 1, "Development of a Geologic Map"). The lava beds in The
Needles are upended and their hardness makes them stand high as
erosion gnaws away faster at the surrounding softer rocks. The pinnacled aspect of the ridge is due to widely spaced joints (cracks). Erosion tends to work along the joints and the beds, etching out blocks,
ribs, and spines (fig. 48). Although much of the lava is in the pillow
form, indicating that it erupted under water (note 25), some of the rock
of The Needles evidently did not reach the bottom of the ocean but
forced its way as a melt between beds of lava already deposited.
These *intrusive sills* (fig. 11) show no pillows. From a distance they
look smoother than the pillow lavas, but on freshly broken surfaces
can be seen to be composed of interlaced, thin, rectangular,
light-colored crystals of feldspar in a black matrix of pyroxene and
altered material. The intruded layers, called diabase where fine-
grained and gabbro where coarse, were insulated from the chill-
ing sea water so that the crystals had time to grow.

In contrast, the quickly cooled pillow basalt is dark green and finely
textured, with few if any visible crystals (fig. 49). Pillows commonly
contain many small, globular, greenish-white structures ranging from
BB to pea size. These are masses of tiny crystals that grew from the
glass that formed a rim on the pillow when the molten glob of lava
came in touch with sea water. Glass, which does not have the or-
dered structure of crystals, never lasts long geologically. Its molecular
components slowly gather together to become crystals, a more stable
form of matter, and the lumps in the basalt are centers of this crystallization.

Gray Wolf River

Downstream from the Slab Camp Spur Trail, the Gray Wolf Trail (47)
clings to the side of a gorge cut in pillow basalt (note 25). At the
bridge crossing (about 2.3 miles from the trail head) cliffs of the bul-
bous pillows rise high in view.

The road to Slab Camp crosses terrain heavily mantled with gravels (48)
and boulders carried down from the north by Cordilleran ice (see note
56). The ice that surrounded the Olympics on the north and east evi-

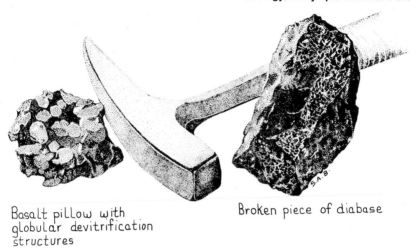

Basalt pillow with
globular devitrification
structures

Broken piece of diabase

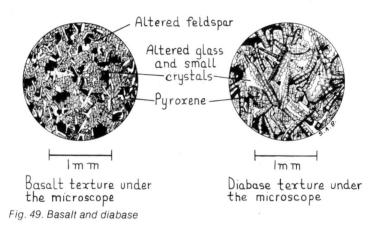

Altered feldspar

Altered glass
and small
crystals

Pyroxene

Basalt texture under
the microscope

Diabase texture under
the microscope

Fig. 49. Basalt and diabase

dently overrode Ned Hill and spilled down Slab Camp Creek into the
Dungeness. Morainal debris left by the ice provides a convenient
ramp for the Slab Camp Spur Trail to descend into the deep canyon.

(49)   The Gray Wolf Trail climbs high after leaving Camp Tony and skirts
rib after rib of rock. Many rock types can be viewed here: green pillow
basalt, disintegrating dark-green and black basaltic sandstone entirely composed of grains of altered lava, red limestone, and shales.
These rocks are now tilted on end, but they indicate a time in the past
when thin lava flows erupted onto the ocean floor. Between eruptions,
sand and mud accumulated.

(50)   Where the trail crosses a rocky landslide are tilted strata of gray
sandstone with coaly layers. When these sands spread across the

ocean floor, some plant material was carried with them and buried. Under pressure, heat has driven off almost all the more gaseous components of the plants—mostly oxygen, hydrogen, and water—and left the carbon.

Cameron Creek

About 4 miles from Three Forks shelter, hikers on the Cameron   (51)
Creek Trail thread their way through a goblin woods haunted by huge green boulders and blocks broken from basalt cliffs above. Up the steep-sided valley to the southeast are two great monoliths of basalt that rise abruptly several hundred feet above the gentler valley floor. The thick but discontinuous layers of basalt here are part of the inner basalt ring (figs. 16, 17).

Just below Grand Pass and northeast of the trail are slabs of sand-   (52)
stone with a bumpy, cobblestonelike surface (fig. 50). These features form when sand is deposited on mud in the ocean bottom and differential settling of the sand into the mud forms bowllike depressions filled with sand. The outcrops here are the bottom of the sandstone bed that has turned almost completely upside down.

Fig. 50. Lumps on the bottom of an upside-down sandstone bed made by differential settling of the sand into mud. The trail to Grand Pass crosses in the middle ground; photo taken southeast of Grand Pass

(53)   The upper part of Cameron Basin is a spectacular example of a
glacial cirque. The glaciers are gone, but their work is seen every-
where in rounded knobs and ridges, smoothed and striated rock
bosses, long lateral moraines running down the valley, and even
fresh, barren terminal moraines draped across the head of the valley
below small patches of ice (fig. 51). A hike to the summit of the ridge
to the east of the basin brings the glaciophile a view of the still-active
Cameron Glaciers, more impressive than the ice relicts above the
main basin (fig. 52).

(54)   At Cameron Pass, the contrast between the precipitous, recently
glaciated north side of the ridge and the gentle south side is particularly conspicuous (fig. 51). If the south side was ever chewed into by
glaciers, it was so long ago that the slow processes of weathering and
rock and soil creep have long since erased the marks.

Deer Park Road*

(55)   The banks of the Deer Park Road clearly reveal the layers, or beds,
of stratified rock. Etched in relief, they form a decorative wall along
the road almost all the way to Deer Park. In many places a regular alternation of light-gray sandstone with dark-gray to black shale is conspicuous; the sandstone beds are hard and angular on the edges and
project farther than the softer, darker, somewhat crumbly shale beds
(see fig. 8). These are rocks formed from sand and mud deposited in
the ocean about 40 to 60 million years ago.

On broken surfaces of the sandstone, the sand grains can easily be
seen. The much finer grains of the black shale are less easily recognized, but the dust produced from a scratched piece of shale is nothing more than dried mud.

At 5.9 miles from the park boundary, where the road begins to
break out into the open, thicker beds of sandstone appear. In fact, the
road becomes quite rough here and just a little steeper, reflecting the
greater difficulty of making a roadcut through these hard beds. High
on the mountainside they form conspicuous ribs. A close look at the
rocks reveals small, rounded pebbles in the sandstone, remnants of
ancient gravels.

(56)   Few viewpoints in the Olympics better allow the imagination to
re-create the time of the Ice Age than does Blue Mountain. To the
north, look out over the plains and waterways to see where the
southern margin of the great Cordilleran ice sheet lay about 15,000
years ago.

On the way up Blue Mountain are large boulders of white granite

* For more information about Deer Park, see "Reading and References."

Fig. 51. Remains of glaciers and their marks in Cameron Basin. The trail crosses Cameron Pass just left of the second ice patch from the right and goes through Lost Pass in the center background. Aerial photo by Austin Post

Fig. 52. Middle Cameron Glacier. Note moraine below glacier and glacially striated rock in foreground

along the roadside (at about 4.9 miles from Deer Park, or 3.0 miles from the park boundary). As far as is known, there is no bedrock of gran e anywhere in the Olympics. Thus these boulders must have been brought here by the Cordilleran ice; boulders of rock types characteristic of the North Cascade Mountains and British Columbia Coast Ranges are common up to elevations of about 3,500 feet all around the north and northeast end of the Olympic Mountains. Visualize the great mass of ice building up and around the dam of the Olympics, one branch of ice flowing out along the Strait of Juan de Fuca to the sea, the other flowing south in the Puget Lowland beyond the city of Olympia, where the ice finally melted as fast as it advanced.

Northwest of Blue Mountain are tree-covered flats on a broad, low divide between the ridge of Blue Mountain and Round Mountain (point X on fig. 53). The edge of the Cordilleran icecap pushed across this divide between the two mountains, for this area is covered with debris left by the icecap, and the slopes leading westward into Maiden Creek are likewise veneered with outwash from the icecap. Even from Blue Mountain rounded outcrops of lava smoothed by the scraping of ice are visible on Round Mountain. Compare the smooth shape of Round Mountain with the jagged, unglaciated cliffs of Mount Angeles.

More granite boulders are scattered throughout Morse Creek valley, its tributaries, and all the other forested valleys west of Blue Mountain, up to an elevation of about 3,500 feet. These boulders, too, might have been carried in by the icecap, although there are no rounded knobs and smooth ridges to indicate that the glacier filled the valley. But with the icecap pressed close around the mountain front, the streams draining the mountain would probably be dammed. If the valley were filled with a lake at the toe of the icecap (fig. 53B), icebergs breaking from the ice, laden with foreign rocks and gravels, could have floated out into the lake where they slowly melted and dumped their load of debris to the bottom far from the edge of the ice. Such ice barges have indeed been observed in present-day northern latitudes where icecaps and glaciers still exist.

The icecap and lake can explain some of the landscape seen today, but two peculiar features are not so easily explained. At the head of Morse Creek, the Cox Valley is broad and flat, singularly different from the narrow valleys of the Morse Creek tributaries. Also, Morse Creek takes an odd bend (see relief map). Flowing for several miles toward the east-northeast and eroding its valley in the relatively soft shale and sandstone, it suddenly takes a sharp swing to the north and cuts through a thick ridge of resistant lava.

Look at the low divide between Round Mountain and Blue Mountain. The divide lines up vertically as well as horizontally with the flat

A.

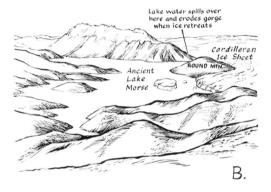

Lake water spills over
here and erodes gorge
when ice retreats

*Cordilleran
Ice Sheet*

ROUND MTN.

Ancient
Lake
Morse

B.

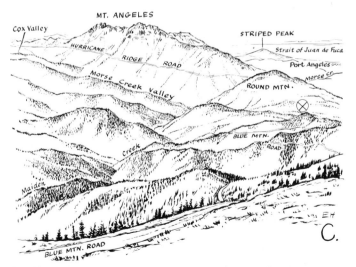

MT. ANGELES

Cox Valley

STRIPED PEAK

Strait of Juan de Fuca

HURRICANE

RIDGE ROAD

Port Angeles

Morse Creek Valley

Morse cr.

ROUND MTN.

BLUE MTN.

Creek

ROAD

Maiden

EH

C.

BLUE MTN. ROAD

*Fig. 53. Ancient Lake Morse and diversion of Morse Creek as seen
from Blue Mountain*

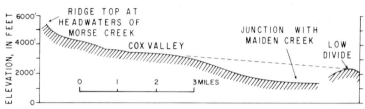

Fig. 54. Profile of Morse Creek

bottom of the Cox Valley (fig. 54), suggesting that the Cox Valley and the low divide are both parts of a once-continuous valley.

A no doubt greatly simplified history might be that before the continental ice filled the lowlands to the north, Morse Creek flowed northeast around the south side of Round Mountain, thence northward over soft sedimentary rocks to the Strait of Juan de Fuca (fig. 53A). Its position here was determined by a long period of erosion, during which time the more resistant lavas of the Mount Angeles–Round Mountain ridge stood out above the valleys eroded in softer rock.

When the icecap grew and advanced to block Morse Creek, a lake was formed that eventually spilled over the lava ridge just west of Round Mountain, perhaps following a course to the strait between the icecap and the mountain front (fig. 53B). When the icecap began to melt away as world climates warmed up, Morse Creek was trapped. It had cut a notch in the hard lava ridge and was already lower than its old course across the low divide. As the creek cut an ever-deepening gorge through the lava, the lake drained away. The land probably rose to some extent as the great weight of ice was removed from it; thus the creek cut well below its ancient course. The Cox Valley and the low divide south of Round Mountain may be remnants of the ancient Morse Creek valley.

(57)   The summit of Maiden Peak, just above the trail from Deer Park to Obstruction Peak, affords an interesting view to the north. The long ridge west of Maiden Lake is composed of thick beds of sandstone, tilted gently to the west so that the top and upper parts of the western side of the ridge are eroded from the sandstone. The blocks of sandstone and sand produced by weathering of the rock make a very porous mantle, through which water from rain and melting snow passes very quickly. The resulting dry soil conditions hinder the growth of the usual fir and hemlock forests of the Olympics; a pine forest thrives on the ridge instead. The hiker who descends to this forest and sandy terrane may well imagine himself in some desert range of the southwest.

Fig. 55. Mastodon. Drawn from a photograph of a painting by Charles R. Knight in the Field Museum of Natural History, Chicago, published in E. B. Branson and W. A. Tarr, Introduction to Geology (New York: McGraw-Hill, 1941), with the permission of the Field Museum and McGraw-Hill Book Co.

Hurricane Ridge Road*

Lake Dawn is a man-made lake scooped in Pleistocene gravels and (58) sands that fill a gentle depression between steep hills of basalt. In similar Pleistocene deposits east of Port Angeles a mastodon skeleton was recovered; the forest-dwelling hairy elephant, relative of the mammoth, once roamed the Olympic Peninsula (fig. 55).

At Lookout Rock, just before the tunnels on the Hurricane Ridge (59) Road (4.2 miles from the Heart o' the Hills entrance station), the traveler is treated to several interesting views. To the northeast are glacier-smoothed hills and, if the day is clear, inlets and low islands carved by the Cordilleran ice sheet are visible. The imagination (or the Park Service display) can conjure up the icecap grinding against the Olympic Mountains (see note 56).

The rock underfoot is basalt, hard and resistant to erosion. By the tunnel, look for the rounded, globular forms of basalt pillows (note 25). To the southeast is the gorge of Morse Creek, cut by the creek after or during the waning of the icecap about 13,000 years ago (note 56).

At several precipitous places along the route (especially at a (60) winding stretch 5.6 miles from the station), the roadcuts along the side of Klahhane Ridge reveal the curiously rounded basalt pillows that indicate that the lava poured out into water. The first bit of hot lava to come in contact with the sea water forms a globule, or pillow, its surface quickly cooled and hardened by the sea water. Often single pillows form and roll down the side of a pile of accumulating volcanic debris to be mixed with the ocean muds. Fragments of pillows and basalt flows, all glued together in a jumble, are even more common than pillows; look for outcrops of this breccia between layers of pillows (fig. 56).

* For more information about Hurricane Ridge geology, see "Reading and References."

Fig. 56. Breccia and pillow basalt on Klahhane Ridge

In some roadcuts are beds of bright red limestone. These colorful rocks are mixtures of mud and submicroscopic skeletons of plankton deposited on the ocean floor between outbursts of lava (note 2); they derive their red colors from iron and manganese leached out of the still-hot volcanic rocks. In these limestones are the tiny shells of one-celled ocean creatures known as Foraminifera, barely discernible with a magnifying lens (fig. 57).

(61)　　At the established viewpoint, look out to see Blue Mountain and terrain once covered by ancient Lake Morse (see note 56).

(62)　　On the trail to Klahhane Ridge, view cliffs of basalt and breccia rubble on the left in the crags of Mount Angeles and to the right find meadowed and hill-covered slopes of sandstone and shale. The hiker here traverses the edge of the upturned lava field where volcanic material was deposited next to quiet accumulations of sand and mud. In one place the trail draws close to cliffs of volcanic breccia and a contorted bed of red limestone. The contortions in these beds may not have resulted from folding during major upheavals in the earth's crust but from sliding and slumping of the partly consolidated strata on the ocean bottom soon after deposition.

(63)　　The ridge crest and trail junction overlooking the Strait of Juan de Fuca provide the most spectacular views of the volcanic rocks of the

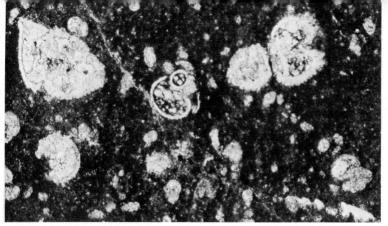

Fig. 57. Fossil Foraminifera in red limestone, enlarged about 27 times

Mount Angeles massif. To the northwest on the shoulder of Mount Angeles, colorful beds of sedimentary rocks, rich in fragments of lava and alternating with beds of volcanic breccia, lean toward the strait (fig. 58). The difference in resistance to weathering of each bed has produced startling ribs and flutes: thick beds of hard volcanic breccia stand up in straight walls; the soft beds of red shale make deep alleys.

When sediments are dumped into the ocean, the largest, heaviest particles or chunks tend to settle to the bottom first, and the smallest last. The result of this simple process can be seen in many outcrops of graded breccia (fig. 59) and tells us which side of a particular bed was originally up. These graded beds show that the lava field of Mount Angeles has not only been tilted up from its original horizontal position but also tipped over

Where the Hurricane Ridge Road cuts the ridge just before the Big Meadow parking lot, slabs of brown, gray, and black slate gleam in

(64)

Fig. 58. Ribs of volcanic breccia east of Mount Angeles

Fig. 59. Graded breccia on Mount Angeles, where originally horizontal bed has been tilted to the vertical and its top now faces to the right

the sun. The slate forms from shale subjected to great pressure, and the rock tends to break into thin splinters.

Fractures develop in two ways to form the splinters (called *pencils*) during the folding of shale under great pressure (fig. 60). The original shale layer is usually made up of many very thin beds and tends to break along or between these beds. The intersection of the closely spaced slate fractures and the original bedding combine to chop the rock into the splinters. In other places pressures exerted from different directions provide several sets of fractures, and their intersection produces splinters.

(65)   The old roadcut along the Hurricane Hill Nature Trail reveals alternating layers of gray sandstone and shale (note 55). These rocks have been strongly folded, and in the outcrop along the trail, the careful observer can trace sandstone beds around the bend of the fold (fig. 61).

(66)   At the last climb of the Hurricane Hill Nature Trail, where the trail begins to zigzag, the mountain stroller reaches the first thin beds of volcanic rock that spilled out across the ocean floor about 55 million years ago. To the left, these lavas make small, rugged cliffs. Rubble weathered from alternating layers of dark volcanic and sedimentary rocks is found all the way up to the gentle meadow on top of Hurri-

Shale layer between sandstone beds

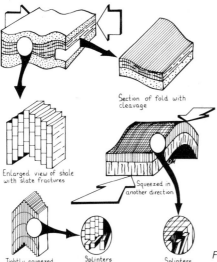

Section of fold with cleavage

Enlarged view of shale with slate fractures

Squeezed in another direction

Tightly squeezed into a fold

Splinters

Splinters

Fig. 60. Formation of slate pencils

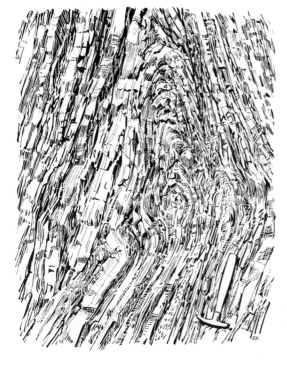

Fig. 61. Fold in sand-stone and slate along Hurricane Hill Nature Trail

cane Hill. Much of the volcanic rock here is sprinkled with white dots, which are small cavities filled with white minerals, mostly calcite and zeolites. The cavities were formed by gas bubbles escaping from the once-molten rock. Later the zeolite and calcite precipitated from mineral solutions in the rock and filled these fossil bubbles. At the very summit, where the old fire lookout used to be, is a ledge of pillow basalt, much like that described along the Hurricane Ridge Road. The main mass of the submarine volcanic field, now tilted on end, can be seen on Mount Angeles across the valley to the east.

(67)   Most geologic processes are of barely comprehensible slowness, but a few are rapid and even catastrophic. Where the Obstruction Peak Road emerges from the forest to balance precariously on a knife-edge saddle before the long hill up to the base of Steeple Rock (about 1 mile from the turnoff), look to the west to see some large blocks of dark rock among the trees on the bumpy terrain (fig. 62). The explorer who wanders down among these monoliths of pillow basalt will find a fantastic jumble of rock walls, blocks, and rubble piles. The scene resembles the ruins of some gigantic city, overgrown and partly hidden by trees. At the base of some huge mountain escarpment we would expect to find such debris fallen from the heights, but there is no such escarpment near here. Up the road, however, is a clue to the ruin in the form of Steeple Rock, an erosion-resistant pinnacle of basalt standing high above the rolling meadows of slate and sandstone. It is part of an uptilted zone of basalt layers, the inner basalt ring, extending many miles to the southeast (see figs. 16, 17). The rocky ruin lies along a projection of the same zone to the northwest. The rubble probably once stood as a pinnacle like Steeple Rock. From the amount of debris, we can surmise that the pinnacle was even larger. The cause of its collapse appears to be landsliding. Slippage of the surface layers of rock, especially in areas of weak slate or shale, is very common in the Olympics (notes 121, 140).

(68)   Along the Obstruction Peak Road are shallow swales paralleling the contour of the hillside; the road follows some of these features (fig. 63). Similar to ridgetop depressions, they are caused by downhill creep of surface rubble and in some places of the bedrock itself (notes 121, 140).

(69)   P. J. Lake, a favorite fishing hole, lies in a glacial cirque. Rivers cannot hollow out large depressions, but glaciers can. The basin containing P. J. Lake is dammed by resistant knobs of basalt, smoothed but not removed by the glacier.

(70)   Much of the Obstruction Peak Road follows a gentle, rolling upland surface. The contrast with the steep, newly eroded valley wall is especially evident where the road leaves the parking lot at Big Meadow

Fig. 62. Collapsed basalt pinnacle on Hurricane Ridge

Fig. 63. Depression along Obstruction Peak Road

and where it switchbacks up past Eagle Point (fig. 64). The meadowed upland surface formed before the last glaciation and probably
at a time when the Olympics were lower and streams had less gravitational energy to carve the mountains (see Chap. 2, "Rising Land and
the Olympic Scene").

(71)   The downhill creep of rock debris is a constant process, and in
early spring the hiker finds strange sags in the high trails, especially
those on the south sides of ridges. The relatively rapid creep of the
top layers of soil and debris has carried the trail downhill. Actual
breaks in the trail are soon mended by the passage of boots.

(72)   In a few places where the Obstruction Peak–Deer Park Trail skirts
the south side of Elk Mountain, the blocks of reddish rock rubble appear to be filled with small, whitish spheres. These spherical structures in the basalt are formed by crystallization of rock glass (see note
46).

(73)   Just south of Obstruction Peak, the trail to Grand Lake leads out
along the crest of the main ridge. At numerous places along this crest,
the terrain is mottled with blocks of black, lichen-covered sandstone
produced by frost weathering. Water expanding when it freezes in
cracks of the rock exerts a tremendous force. The continual freezing,
thawing, and refreezing of water breaks up exposed bedrock and produces the rocky rubble found on many high ridges in the Olympics.

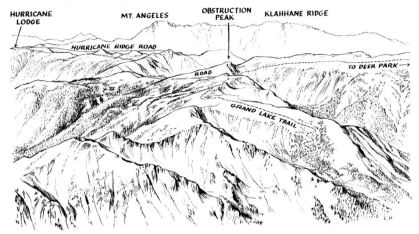

Fig. 64. View of Hurricane Ridge from the south. The old upland surface can be seen to the left near the ridge crest, steep valley sides below, newly cut glacial cirques on the right

The view across the valley of Grand Lake from the trail gives the   (74)
hiker the impression of very disordered geologic structure (fig. 65).
Cliffs and pinnacles of resistant sandstone beds stand out patchily on
the ridge southeast of Grand Lake. In a normal succession of sandstone and shale, even if uptilted or folded, the pattern of the bedding
revealed by erosion would probably be discernible in a view like this.
Here, however, the continuity of strata has been disrupted. The rocks
are folded, and the beds and folds have also been pulled apart along
many small faults. Isolated masses of resistant sandstone beds remain in the broken slate (see notes 105, 125, 137).

Fig. 65. View across Grand Valley, showing masses of resistant sandstone in a mush of slate

Fig. 66. Complex fold
on west side of Grand
Valley, with Moose Lake
on right

(75)   On the hillside west of the Grand Pass Trail are some spectacular examples of folded rocks (see fig. 66).

(76)   The hiker on the ridges west of Grand Valley has a good chance of finding some fossils. Look for, *but do not collect or disturb,* flattened, tubelike structures in slates—mere imprints on bedding surfaces—or sand-filled tubes in slate or fine-grained sandstone (fig. 67). They may be straight or curved and range from less than an inch to several inches long. The structures are worm tubes or borings, and their abundance in much of the Olympic core rock indicates the ancient sea bottom was teeming with creatures. Unfortunately, we do not know much more about them because the soft-bodied worms themselves almost never survive as fossils.

(77)   The vast cirque area of the upper Lillian nicely illustrates the alpine evolutionary sequence from glacier barrens to alpine meadow to Hudsonian forest (fig. 68). The ancestral Lillian Glacier shaped this land, and a tiny glacier, reborn about 2,500 years ago, still clings to the northern cirque walls. Below the ice, barren moraine debris fills in around glacier-smoothed rock bosses of sandstone and slate. The barrenness testifies to the recency of the latest glacier retreat. The glacier-carved basin of Lake Lillian is still bleak, but meadows are forming, and flowers and grasses abound just downhill from the lake. Farther down is meadow and forest. A large, flat meadow below Lake Lillian was probably a glacial lake, now completely filled and

someday to be forest-covered. Deep forest fills the valley below. Many of the lower Olympic valleys must have gone through this same cycle after the Pleistocene glaciers retreated and the climate warmed (note 120).

Elwha River

Just south of Highway 101 the Elwha River Road leaves an open, peripheral valley on the flanks of the Olympics and follows the river through a notch in a thick mass of basalt, the north arm of the Olympic basaltic horseshoe (fig. 17). Outcrops along the road where it is close to the river reveal the rounded pillow forms of submarine lava (notes 25, 26, 60). Views to the west as the road progresses up the valley reveal cliffs and rock promontories of the basalt on the shoulder of Mount Baldy. The steep, hard masses of basalt discourage tree growth, and the cliffs are large enough to show through the trees around them. Where the valley walls pass into shales and sandstones, smaller cliffs in these softer rocks are completely hidden in the trees. (78)

Lakes filled many of the larger valleys on the north and east side of the Olympics when they were dammed by the Cordilleran ice sheet (see note 56). Deposits of sand, gravel, and clay that washed into these lakes may still be found along the valley sides. Clay deposits, like the one along the road a short distance below Lake Mills, are particularly prone to landslide and creep, especially when wet. The continued movement of the clay downhill prevents many conifers from growing, tilts those that do, and continually destroys the road. A retaining wall attempts to prevent the material of the slope above from flowing onto the road. (79)

Fig. 67. Worm burrows that have been filled with fine sediment, then flattened and distorted

Fig. 68. From glacial barrens to meadow and forest. View is west over upper Lillian basin, with Lillian Glacier at left and Lake Lillian in center; in middle background across the Elwha valley is Dodger Point and, behind it, Mount Carrie. Aerial photo by Austin Post

(80)   Observation Point is a good place to view the Elwha Valley. The broad U-shaped valley as seen looking upriver was carved by a glacier; but the subsequent history of the valley includes river erosion, sedimentation in ancient Lake Elwha (dammed by the Cordilleran ice sheet), and further erosion (fig. 69). The broad valley floor seen upriver from Observation Point is formed by remnants of the glacier-carved floor and gravel terraces of the lake stage, most noticeable as the tree-covered bench immediately across Boulder Creek west of Lake Mills.

Observation Point is also a good spot to examine mica-rich sandstone beds. Freshly broken surfaces of the gray rock sparkle with small mica flakes carried into the ocean between 40 and 50 million years ago. The mica is mostly clear and soft; you can chip out tiny, thin flakes with a knife.

(81)   Where the road swings into the steep canyon of Crystal Creek, look for black, contorted shales with broken slabs and blocks of gray sandstone and limestone. This is a small exposure of the wide Calawah fault zone (see note 83).

(82)   The hot waters of Olympic Hot Springs pour out of a brushy, rubbly hillside on the east side of Boulder Creek. The waters were once gathered behind numerous dikes and in tubs, then piped into a large swimming pool. At one time the springs were the site of a unique

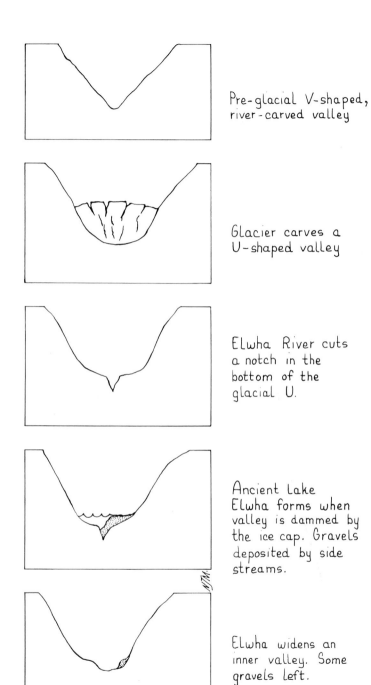

Pre-glacial V-shaped, river-carved valley

Glacier carves a U-shaped valley

Elwha River cuts a notch in the bottom of the glacial U.

Ancient Lake Elwha forms when valley is dammed by the ice cap. Gravels deposited by side streams.

Elwha widens an inner valley. Some gravels left.

Fig. 69. Simplified history of the Elwha River valley in cross section

wilderness spa, reached only by trail. About 1923, E. B. Webster
wrote of a traveler's arrival at the springs:

> All day, if he has slowly sauntered the ten mile trail, he has been enjoying
> the pure mountain air, laden with the perfume of flowers and the scent of the
> balsam. But now there drifts down the trail a faint, elusive odor of an entirely
> different character, something like that of ancient henfruit. Only a mere whiff at
> first, and for a moment he hesitates. Then another slight breeze ripples the fo-
> liage and he has it placed. The springs are not far distant now and presently
> he will note the thin cloud of vapor arising, for here are some twenty-odd sep-
> arate springs, some hot, some hotter, and some—hot!
>
> Hung on a narrow shelf on the mountainside, over a canyon threaded by the
> roaming Boulder in its express train speed to the Elwha and the sea, the hotel
> presently comes into view. A hundred-foot-high bridge is to be crossed before
> one reaches the hotel, the bath houses, the open air pools and the rows of
> tents and tent-houses where comfortable beds and abundant fresh air insure a
> night of deep, unbroken slumber.
>
> A number of guests are diving and swimming in the larger pool; others are
> strolling on the board walk, while still others, including those whose appetites
> have been sharpened by a day spent on the mountain ridges or in casting in
> the pools below the waterfalls, are patiently awaiting the ever welcome sound
> of the dinnerbell.

Although the springs have fallen into disuse, their waters are as hot
as ever. The waters of Olympic Hot Springs appear to be little different
chemically from plain surface water that has been heated and passed
through the rocks. We cannot say exactly why the springs are hot, but
they do occur on the Calawah fault zone (note 83), and the broken
rock in the fault zone might provide channels for water to circulate
deep into the hotter interior of the earth (fig. 70).

(83)   Crystal Ridge, reached by the Old Crystal Ridge Trail from the
Olympic Hot Springs Campground, is carved from sheared rocks of a
wide fault zone. This fault, named the Calawah fault, has been traced
westward more than 25 miles from Crystal Ridge to the Soleduck
River near Sappho and probably stretches westward under glacial
gravels to the ocean. Eastward it curves around roughly parallel to the
Olympic basaltic horseshoe and branches out into other faults and
slate belts (see notes 9, 35, 172, and figs. 16, 17). The Calawah fault
and related faults are not known to be active now, but they were active
millions of years ago when all the rocks now seen were buried deeply
under the surface.

On Crystal Ridge the observer will see slaty shale, mostly weathered into small chips. Scattered throughout are broken beds and
pieces of sandstone, limestone, and conglomerate—hard fragments
milled along in the softer rocks.

The Calawah fault separates sandstone (seen as cliffs and ribs on
the lower south side of the ridge and beyond to the south near Mount

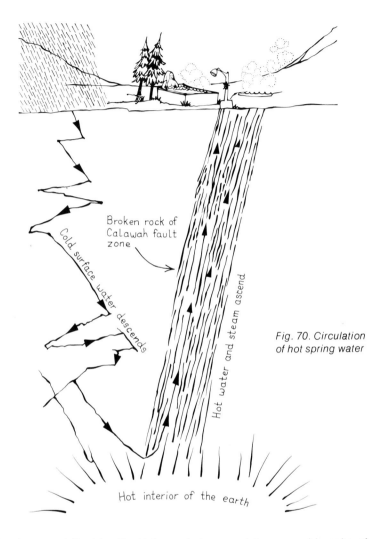

Broken rock of
Calawah fault
zone

Cold surface water descends

Hot water and steam ascend

*Fig. 70. Circulation
of hot spring water*

Hot interior of the earth

Appleton and Boulder Peak) from shales, sandstones, and basalts of
the basaltic horseshoe (seen on Happy Lake Ridge to the north).

Although most of the Olympic core rocks are highly folded, in only (84)
a few places are large folds well displayed. One such fold, a syncline,
(see fig. 71 and also note 75) is easily seen on Mount Appleton as
viewed from Appleton Pass.

Oyster Lake is part of a ridgetop depression, a gap along the ridge (85)
left by slumping or sliding of the rock toward the valley (note 121).

Near the end of the Whiskey Bend road, look across Lake Mills to (86)

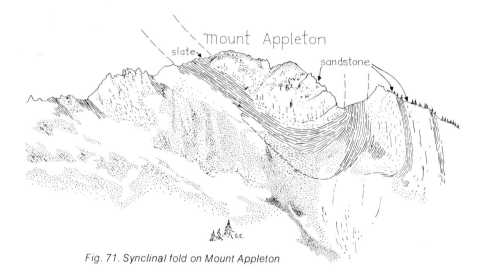

Fig. 71. Synclinal fold on Mount Appleton

see high bluffs of bedded gravels, probable remnants of the deltas of Boulder and Cat creeks, which were built into ancient Lake Elwha (note 80).

(87)   The Elwha River Trail reaches a high point just above the Elk Overlook. It makes this climb to avoid the steep sandstone cliffs of Rica Canyon. Just below the overlook, the entrance to this gorge was called the Goblin's Gate by the 1889 *Press* Expedition (the gate may be reached via the Krause Bottom Trail). Above the resistant sandstone beds, the Elwha has opened up the small flat-floored Geyser Valley (including Krause Bottom—see notes 80, 88).

(88)   Geyser Valley was named for geysers that the *Press* Expedition thought they heard but never found. The valley, like many along the Elwha River, lies between two stretches of rock-walled gorge, Rica Canyon below and the Grand Canyon above. The gorges are cut in hard sandstone with some slates, and although exposures of rock are rare in the wider valley, it has probably been cut in less resistant slates. Similar but smaller valleys between gorges may be found along the trail upriver (see also note 80).

(89)   In the very recent geologic past, a large mass of rocks from the western side of the canyon slid into the Elwha and dammed the river. A lake formed behind the debris, but in 1967 the dam broke and the lake water washed out the trail bridge and deposited gravels in the bottom lands below. The flood damage is easily seen on the Long Ridge Trail where it approaches the new steel bridge. The cause of the landslide was twofold. First, the rocks on the steep northeast side

of the ridge leading north from Dodger Point were bent over and fractured by creep similar to that which produces ridgetop depressions (see notes 121, 140); then the river cut at the base of the unstable slope until the unsupported, loosened rock broke loose. This interplay of gravity and running water is one of the most important ways by which the Olympics are eroded.

After leaving Idaho Creek, the Elwha Trail traverses a glacier-cut (90) terrace mantled with gravels (note 80). The dry aspect of the forest and its slow recovery from a forest fire are probably due to the porous gravel, which lets rainwater drain quickly away.

Just beyond the Lillian River, the Elwha Trail climbs high on the (91) valley side to avoid the steep walls of the Grand Canyon of the Elwha (also known as Convulsion Canyon), a steep-walled gorge cut in a sinuous pattern. To view this gorge best, leave the trail where there is an obvious drop-off below and scramble down to the edge of a 600-foot rubble slide. The peculiar knob of rock that seems to block the straight course of the river, throwing it into a U-bend, is a glacially carved boss of resistant pillow basalt. Most likely the river was diverted around the knob by ice or moraine at a much higher erosional level and has since entrenched itself in this convolute gorge (see note 166). Forced to one side of the valley, the river undercuts the slope below the trail, creating the abyss below the viewpoint.

Beyond Mary Falls Shelter the trail climbs to a bench cut in bedrock (92) and mantled with creeping clay deposits at Wildrose Creek. The bench is another remnant of the old glacier-cut valley bottom, and the clay is the sediment of an Elwha valley lake (notes 79, 80).

Just below Remanns Cabin, the river has cut through steeply dip- (93) ping black slate and phyllite that are conspicuously displayed along the trail (note 64).

The Semple Plateau, named by the *Press* exploring party, is a (94) strikingly flat series of terraces. The terraces are mainly cut in bedrock slates and sandstone but are thickly mantled with gravels, which probably represent the outwash delta of the Goldie River where it entered ancient Lake Elwha (note 80). The Goldie River cuts an impressive gorge through these terraces.

The Elwha River appears to have widened the Press Valley by (95) side-cutting as it has other inner valleys downstream (note 88). A difference in rock hardness between the Press Valley bottom and the gorges cut in interbedded sandstone and slate downriver is not readily apparent, but the rock underlying the valley probably contains a greater number of slate beds.

Just above the bridge over Hayes River, where the trail swings south- (96) west along a steep, brushy hillside and before it turns south above the Elwha once more, the hiker can find outcrops of varved clay and silt. These remarkably even-bedded clays and silts represent lake deposits. The first layer (or *varve*) of silt may represent one year's spring flood of silt from the rivers; the adjoining clay layer, the winter's slow accumulation of mud settling out of undisturbed lake water. These varved clays give evidence of a lake that may have been either a local pond caught between the Elwha valley glacier and the valley side or ancient Lake Elwha, which filled the Elwha valley when the mouth was dammed by Cordilleran ice about 15,000 years ago (note 56). Other evidence for Lake Elwha can be found in cobbles and pebbles of granite and metamorphic rocks as far upriver as Godkin Creek. Such rock, found nowhere in the Olympic bedrock, could only have traveled on icebergs floating up the lake.

(97)　　　The trail wanders from one rock-cut river terrace to another before it reaches the river bars at Camp Wilder (note 80).

(98)　　　Mount Dana is an interesting geologic puzzle. Usually much of the shape of the land can be related to the types of rock underlying it

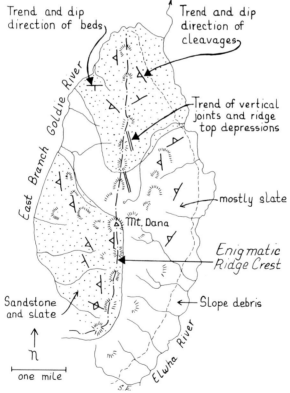

Fig. 72. The straight ridge of Mount Dana: a geologic puzzle

*Fig. 73. Quartz crystals (very
well formed), which range from
1/16 to 2 inches long*

and the erosional processes degrading it. The main north-south ridge
that culminates in the summit of Mount Dana has a remarkably straight
ridgetop trend that seems to belie random natural processes. Commonly, joints or faults make straight landscape features (notes 32,
104); but Mount Dana's straightness does not seem to be related to
any geologic structure, rock type, or erosional process (fig. 72).
We need to know more about Mount Dana to solve this puzzle.

The mountain climber on the north ridge of Mount Norton will find   (99)
clean cliffs of coarse-grained sandstone and small-pebble conglomerate. Here the pressure and squeezing of the conglomerate deep in
the earth has deformed the pebbles and sand grains to elongate rods;
similar rocks occur on Mounts Anderson and White to the southeast.

Along the ridge of Crystal Peak are veins of quartz rich in cavities   (100)
studded with well-formed quartz crystals (fig. 73). Crystals develop
the best shape where they can grow in a free environment such as in
a liquid or gas. These crystals apparently grew in pockets of liquid
(silicon solutions) that later drained, leaving the crystals free of en-
closing minerals. Seek out and examine these crystals, but *do not disturb or collect them.*

The hiker approaching Low Divide from the Elwha valley leaves a   (101)
terrane composed predominantly of slate and enters a rugged terrane
characterized by massive sandstone beds (fig. 17). Just below Low
Divide, the trail, effectively barred from ascent by a thick bed of sandstone, follows a narrow ledge below the sandstone cliff before finding
a break through it.

Low Divide is a fine example of a pass carved by glaciers. It is   (102)
U-shaped, and the viewer can easily imagine ice from the Elwha Gla-
cier spilling through the divide and merging with ice flowing down the
North Fork of the Quinault.

Climb to Martins Lakes or the 5,468-foot peak to the southeast to   (103)
see a fine view of geologic contrasts. To the northeast are long,
even-crested ridges—Wilder, Dana, Norton, and Claywood—only

rarely punctuated with craggy sandstone summits. The mountains at the viewpoint and those forming a belt from the northwest to the southeast, including Queets, Meany, Seattle, Christie, Taylor, and Muncaster, are very rugged and abrupt. The contrast is heightened by climate: a viewer looking northeast in the Olympics sees mostly the south sides of the peaks, which are characteristically less rugged than the snow and ice-laden north sides that he sees when he faces southwest (compare the two sides of the ridge in fig. 51). Bedrock is even more important in forming this contrast. The terrane in immediate view to the northeast is mostly weak slate with some beds of sandstone; the belt to the southwest is mostly thick beds of hard sandstone that stand up in spectacular cliffs (fig. 16). The thick beds of sandstone characteristically break off in large blocks, as seen below along the Martins Lakes Trail.

Some of this sandstone, exposed on Mount Christie, has a greenish cast owing to the metamorphic minerals chlorite and pumpellyite, which formed in the rock when it was buried deep in the earth (note 136).

(104) After many miles of winding trail and river, hikers may be amazed at the thin, straight uppermost Elwha valley and the Elwha snow finger leading to Dodwell-Rixon Pass (fig. 74). Although the snow hides much of the rock, the valley is probably eroded out along a fault. Faults are breaks formed by the movement of two blocks of rock in the earth's crust. Many faults are straight, and the ground-up rock along the fault is more easily eroded than the rock to either side; a straight gully or valley commonly develops along the weakened rock, and the valley's straightness gives the geologists a clue to the structure.

(105) The Queets basin is known for its spreading meadows, shiny tarns, and cascading brooks. The upper basin also contains rocks that demonstrate intense deformation of the earth's crust.

The recently deglaciated outcrops of the basin beautifully display broken slabs of sandstone beds (mostly sheared and recrystallized to semischist) from inches to hundreds of feet long, scattered throughout

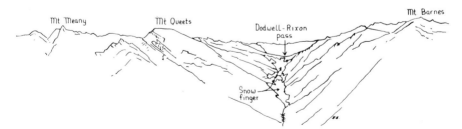

Fig. 74. Elwha snow finger: a straight gully eroded along a fault

Fig. 75. Disrupted sandstone beds (light-colored ovoids) between streaky layers of phyllite and semischist, south of Mount Barnes. Glacial striations trend directly away from viewer

crushed shale, now recrystallized to slate and phyllite (fig. 75). These crushed rocks (and many others like them in the Olympics) probably represent deep-ocean sedimentary rocks that have been swept against and under the edge of the continent as the oceanic floor moved under northwestern America (see Chap. 1, "Colliding Plates and Olympic Rocks").

Long Ridge shows the contrast between geologic processes (106) working on the south and southwest sides of the Olympic Mountains and those on the north and northeast (note 103). Most of the trail climbs slowly up the gentle west side; there are a few views of the precipitous east side from lower stretches of the trail, but the view from the top of Dodger Point is the most impressive (see also fig. 68). On the west side of Long Ridge, much of the upper part is a remnant of an old upland surface where weathering processes and slow creep have had a long time to work (see Chap. 2, "Rising Land and the Olympic Scene"). At places the trail skirts steeper slopes where Long Creek and its side streams are eroding the old surface. The east side of the ridge was cleanly and steeply carved by the Elwha Glacier, which was larger and more active than the glacier in Long Creek. Small glaciers have sharpened the ridge's crest.

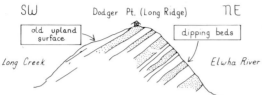

Fig. 76. Simplified cross section of Dodger Point

The general westward dip of the bedrock may contribute even more to the ridge's asymmetrical shape. The more gentle western side roughly parallels the bedding and cleavage, while the eastern side parallels steep joints or, at best, the jagged face of broken-off beds (fig. 76).

(107)　　Near the end of the abandoned trail out to Ludden Peak, the mountain traveler traverses steep ribs of slate. The black rock breaks along minutely spaced planes, the slaty cleavage (fig. 60).

(108)　　Ludden Peak is composed of deformed conglomerate. Pebbly gravel that once came to rest on the ocean bottom has become rock through long burial and recrystallization of minerals between the pebbles. It has also been so strongly squeezed that all the pebbles are elongated. Some broken surfaces of the rock display these pebbles. The resistance to erosion of the hard, tightly knit conglomerate lets it remain as a rugged peak on the ridge of more easily eroded slates.

(109)　　The summit plateau of Mount Ferry is cut by deep cracks, evidence that the weak rocks cannot hold themselves up under the pull of gravity. Cracks such as these commonly develop into long depressions paralleling ridgetops. Several large ridgetop depressions, partly filled with snow and rubble, can be seen on the south end of the Bailey Range (southwest of Mount Ferry) north of Bear Pass (fig. 77). Evidently the rocks, with steeply standing bedding or cleavage, cannot withstand the pull of gravity on the steep flanks of the ridge, and the layered rocks bend over downhill, leaving a gap along the ridgetop (notes 121, 140).

(110)　　The ridge leading north from Mount Ferry is a rock that weathers red-brown, in contrast to the browns and grays around it. The ridge is a rib of greenstone (metamorphosed basalt), and the red color of the surface is due to the oxidation of iron minerals.

(111)　　Although sandstone beds are smashed and shale is smeared out into slate in the Cream Lake basin area, many undisrupted, glacier-smoothed blocks in the deformed terrane reveal relict features of the rock's sedimentary history (fig. 78). Compare this rock, with its beautifully preserved sedimentary features, to a similar rock (fig. 75) that has been sheared and recrystallized.

Fig. 77. Ridgetop depressions filled with snow in middle background. View is north along the south end of Bailey Range, with Mount Childs, center, and Mount Pulitzer on the right skyline

Fig. 78. Sedimentary features in sandstone and slate of Cream Lake basin

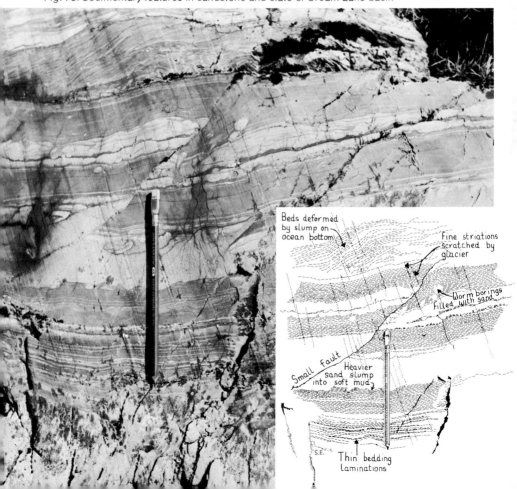

Lake Crescent

(112)   The beauty and spectacular setting of Lake Crescent have long inspired people to speculate about its origin. In 1909, Albert Reagan guessed that Lakes Crescent and Sutherland were downwarps in the earth's crust. There is little evidence for this speculation, but interestingly enough Reagan also reported a Quileute Indian legend about the lake that gives the following explanation:

> Once, in the valley which the lake now occupies, our people and Clallams were having a big battle. For two days the people killed each other. Then Mount Stormking became enraged. You know the mountain that overlooks the north end of the lake from the east. Well, Mount Stormking got angry (all things on earth were living beings then) and he took a great piece of rock from his crest and hurled it down into the valley, killing all who were fighting and at the same time damming the stream with the great rock, so that it has been as it is now ever since, and no Indian has gone near the place since that day.

The Indians appear to be right. Parts of the basins now holding the lakes were scoured out by the Cordilleran ice and once drained eastward via Indian Creek; but according to the work of Robert Loney, writing as a student more than twenty years ago, the low divide between Lakes Sutherland and Crescent is a great landslide mass. Modern studies indicate that most of the debris now seen on the surface came from the north, not from Mount Storm King, as was described in the legend. The sides of the valley probably fell (see note 115) soon after the Cordilleran ice retreated from the valley, and the rock debris partly filled ancient Lake Crescent, splitting it into the two lakes seen today (fig. 79). The waters of new Lake Crescent then spilled north out the Lyre River. Did Indians witness this cataclysmic event, or were they very astute observers of the natural environment?

(113)   Where Highway 101 winds between Lake Crescent and the basalt massif of Mount Storm King, look to the cliffs on the south to see many excellent exposures of pillow basalt (notes 25, 26, 60), especially where the road cuts across the lava beds standing vertically at Sledge Hammer Point.

(114)   Most of the human settlement along Lake Crescent is built on deltas, sloping piles of gravel and silt left by side streams when they enter the lake. The largest delta, at the mouth of Barnes Creek, provides level terrain for park facilities and guest housing.

(115)   Although there is considerable evidence for relatively recent uplift and warping of the earth's crust in the Olympic region, such as raised wavecut terraces near the ocean and folded glacial gravels, there are only a few places where direct evidence of recent faulting can be found. Barnes Creek valley offers some evidence of geologically recent faulting, but the story is equivocal. On the northeast side of the

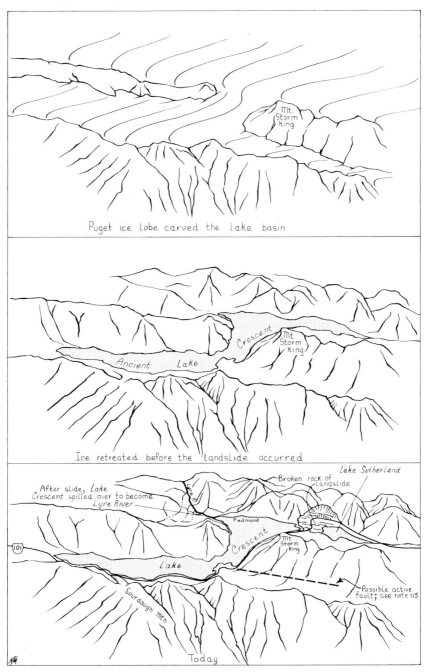

Fig. 79. The formation of Lake Crescent

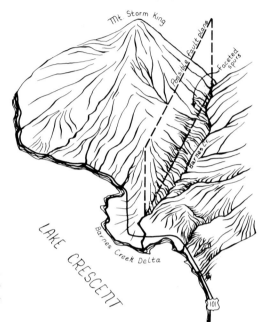

Fig. 80. Faceted spurs
suggest recent fault
movement in Barnes
Creek

valley, steep basalt spur ridges rise up side by side to conspicuous
shoulders. Above these shoulders are faceted spurs (fig. 80). A glacier can cut off the toes of side spurs in a valley and produce the
same effect (note 162), but usually the ice cuts at the level of the
valley bottom. The facets of the spurs here are aligned in a common
plane, and the rock of the shoulder below is broken and crushed. The
evidence indicates that the north side moved up relative to the south
side and recently enough so that erosion has not removed the even-
ness of the fault plane (that is, the facets). Could it be that earthquakes associated with this faulting shook down the landslide that
divided ancient Lake Crescent? (See note 112.) As yet no other evidence has been found to support this speculative question.

(116)   The long natural ramp ascended by Highway 101 as it leaves Lake
Crescent at Fairholm is built mostly of glacial deposits but is underlain
by a bedrock divide, the west edge of the deep glacier-carved basin
bearing Lake Crescent.

Soleduck River

(117)   Figure 81 illustrates how the courses of the north and south forks of
the Soleduck River may have been altered by the Soleduck Valley
Glacier.

(118)   Sol Duc Hot Springs have been of interest to man since early times.

In the 1880s, according to the Port Angeles *Daily News,* a pioneer hunter named Theodore Moritz befriended an Indian with a broken leg and took care of him until his leg had healed. The Indian, in gratitude, told Moritz of the hot springs, explaining that his people had long used the water for treatment of various ailments.

Sol Duc Hot Springs has been in operation as a resort bath since 1912. The origin of the hot water is unknown, but like the waters of Olympic Hot Springs, it is chemically most like surface water, not water from deep in the earth. The springs do not lie directly on the nearby major Calawah fault zone, as do Olympic Hot Springs (notes 82, 83), but surface water might well circulate deep in the earth along the same fault zone and find its way to the surface via joints in thick sandstone beds adjacent to the fault.

The waters of Soleduck Falls have cut a deep slot in massive sandstone beds. The bedding of the sandstone here stands vertically, and the river parallels the bedding, for it can erode more easily along the beds—probably along shaly layers—than across. **(119)**

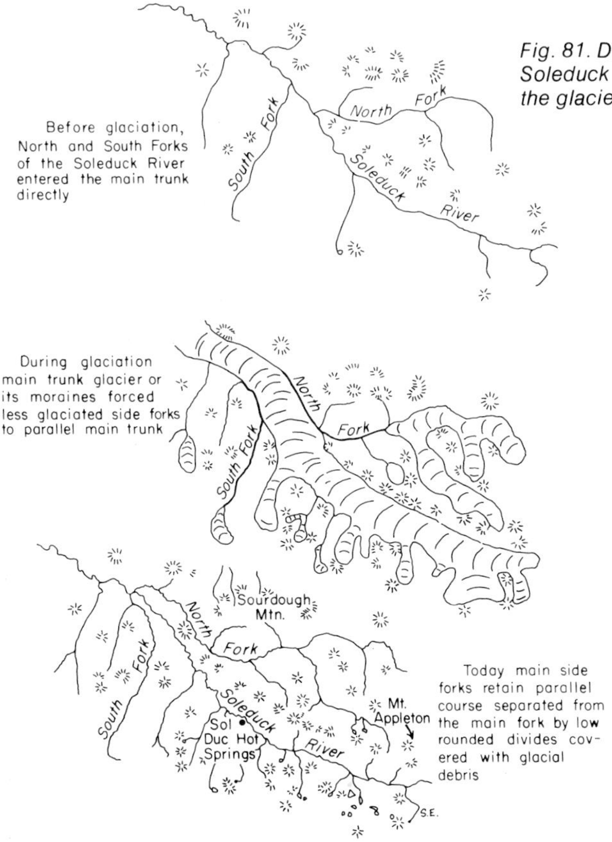

Fig. 81. Diversion of the Soleduck tributaries by the glacier

Before glaciation, North and South Forks of the Soleduck River entered the main trunk directly

During glaciation main trunk glacier or its moraines forced less glaciated side forks to parallel main trunk

Today main side forks retain parallel course separated from the main fork by low rounded divides covered with glacial debris

Fig. 82. Small meadows develop on the site of former glacial tarns above Deer Lake, looking north

(120)   Above Deer Lake are small, flat, sometimes marshy meadows (fig. 82). These are the sites of former glacial lakes called *tarns,* and they illustrate a stage in the evolution of alpine landscape. Many small alpine glaciers on the north side of Olympic ridges either retreated or wasted in place to leave small tarns in basins carved by the glaciers or dammed by moraine. The tarns never last long by geologic standards of time, for silt and sand washed in by streams, coarser debris brought down by winter avalanches, and rockfalls soon begin to fill them up. In the Olympics another process promotes their demise even sooner. When vegetation becomes established in the ice-free basin, sphagnum moss begins to grow out into the lake. Eventually it can cover the lake, long before the lake is completely silted in, and a marshy bog is born. If you walk on the edge of some lakes you can feel the floating moss give way.

      When the silting is complete, the high meadow or marsh may persist a long time if the ground is still too wet to allow tree growth. Avalanche snow or only deep winter snow may help prevent the conifers from taking seed. As the meadow fills and dries, the conifers take

Oversteepened valley side begins to creep valleyward   Beds are finally tipped over and broken

Fig. 83. Large ridgetop depression at the head of the Bogachiel River

root. All elements of this progression can be seen in the high Olympic valleys (notes 77, 168).

At the head of the Bogachiel River, the Bogachiel Peak Trail de-   (121)
tours around a peculiar depression filled with a jumble of sandstone blocks. Gravity produced this ridgetop depression by pulling the steeply dipping beds of sandstone away from the Bogachiel Peak ridge and down into the cirque below (fig. 83). It is likely that this sort of ridge failure, quite common throughout the Olympics, came shortly after alpine valley glaciers melted away, removing their support from the oversteepened valley walls (notes 109, 140).

(122)   Seven Lakes Basin contains more lakes than most areas of the
Olympics. The ridge is mostly underlain by thick beds of sandstone.
The sandstone has coarsely spaced joints, and although it is fairly resistant to much erosion by abrasion, it is particularly susceptible to
glacial plucking, whereby large blocks are pulled out by the ice. The
ridge is oriented east-west, favoring healthy glacier growth on the
north side during glacial times. The resistant sandstone does not now
produce abundant fine silts to wash in and fill the lakes. Because of its
favorable orientation to the north, the basin has probably been completely free of glaciers only for a short time and thus the lakes have
persisted longer than in the lower basins (note 120).

(123)   Where the Bogachiel Peak Trail leaves timberline trees and begins
a zigzag up the steep meadow toward the peak, thick beds of sandstone and conglomerate crop out. The sandstone beds were deposited horizontally at the bottom of the sea, and the rocks preserve features indicative of their watery birthplace many millions of years ago.
Commonly, when a mass of sand and silt is introduced into the deep
ocean as thick, turbid slurry, the coarse material settles first and the
fine material later. Thus, layers of the sandstone are coarse-grained at
the bottom, grading to finer sands toward the top (fig. 84). The geologist can use this feature to find the original top of the bed—no easy
task in the Olympics, where many beds are upside down.

(124)   Many ridges are marked by ridgetop depressions (notes 121, 140),
and the gentle swales and humps by these features provide a natural
route for the High Divide Trail. The depressions are particularly noticeable where the trail leaves the lowest saddle at the head of Cat
Creek and climbs toward the steep side of Cat Peak.

(125)   Some features of the earth's surface arise from large-scale geologic
processes such as movements of large segments of the earth's crust,
as in plate tectonics (see Chapter 1, "Forces in the Earth: Plate Tectonics"), and mountain building. Such forces have raised ocean-bottom sands into the vertical buttresses that now form Cat Peak; much of
the bedding in these sandstones is almost turned on end. But the up-
turning of the beds here is only a part of a much larger disturbance
that broke loose a sandstone slab that stretches from the lower Soleduck River through Seven Lakes Basin, Cat Peak, and Mount Mercury to Queets Basin (fig. 16 and relief map). Much of the Olympic terrane shows evidence of folding and large-scale tearing apart, apparently when the sediments of the ocean bottom were ground against
the edge of the North American continent. The sandstone slab here
evidently was strong; it resisted folding and smashing and survived
nearly intact between wide belts of strongly disrupted slate and sandstone. This is perhaps the largest fragment in a whole spectrum of

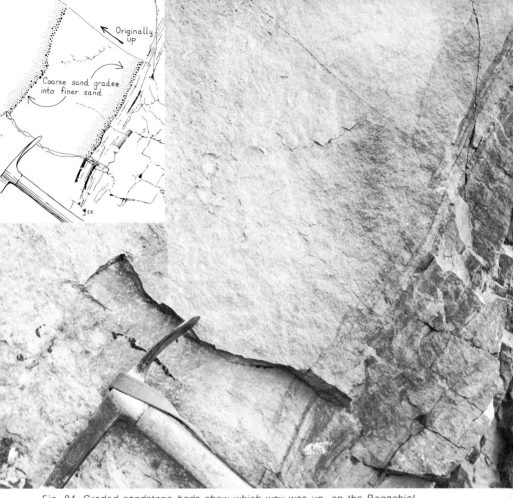

Fig. 84. Graded sandstone beds show which way was up, on the Bogachiel Peak Trail

fragments ranging in size from this twenty-mile-long slab down to broken sand grains, all of which can be found in the disrupted rocks of the Olympics (notes 35, 105, 137).

A glacier-carved valley may contain many waterfalls because the glaciers erode the valley in a rough fashion, commonly plucking out large blocks of rock or grinding down unevenly where side glaciers add more ice to the glacier stream (note 133). Rocks of varying hardness underlying the valley also help to form waterfalls. Along the upper Soleduck River Trail the many steep, glacier-carved steps in the valley profile, commonly partly supported by resistant sandstone beds, give the streams ample opportunity to cascade and fall. (126)

At about 2.5 miles, the Soleduck Trail winds around small forested hummocks interspersed among wooded meadows. The valley bottom here is partly glacier-carved bedrock knobs and partly morainal piles. (127)

Western Approaches

Bogachiel River

(128) Indian Pass is a strange low divide between the South Fork of the Calawah River and the Bogachiel River. The hiker crossing the pass will be mainly impressed by the narrow defile cut by the trail through thick timber, but he cannot help noticing the level terrain, abruptly ending where the trail descends gravelly banks into the Bogachiel. The elevation of the level pass is almost the same as that of thick gravel terraces along the Bogachiel, and it may be underlain by the same gravels as the terraces. Just north of the pass the South Fork of the Calawah changes course from southwest to northwest, and on the map looks as if its upper course were once tributary to the Bogachiel. The gravel terraces here probably formed in glacial times when the whole west side of the Olympics was flooded with gravels (see Chap. 2, "Glaciers: The Heavier Hand"). The South Fork of the Calawah, which apparently originally drained into the Bogachiel, may have been forced by the piled-up gravels, or possibly by the Bogachiel glacier itself, to spill over into a northern drainage channel leading to the Calawah.

(129) Between Flapjack Shelter and Sunday Creek, the Bogachiel River Trail appears to hug the Bogachiel River bank, but there is little water in the gravelly channel. This is not the main river, but a cutoff meander, an old bend of the river that was abandoned when the river, probably in flood, found an easier path across the bend and cut a new channel (fig. 85). At flood time this old channel may be used again, but it will eventually be completely abandoned. This process, whereby the river makes bends and then eventually abandons them, widens the flat river valley.

(130) About four miles above Flapjack Shelter the trail climbs along the side of a gorge to the footbridge over the North Fork. The gorge is cut in thick beds of sandstone on which a terrace developed. The river or a glacier may have cut the surface of this terrace at the same time the gravel of the terraces in the Indian Pass region was deposited (note 128).

Hoh River

The Hoh River valley once bore a mighty glacier that reached clear   (131)
to the sea. Fed by heavy snowfall high in the Mount Olympus region,
the ice carved out the broad U-shaped valley as it flowed westward.
Views of this wide valley and that of the South Fork from the river-
bank gravels just within the national park (13 miles from U.S. 101)
impress the traveler with the cutting power of the glacier.

In the Hoh campground area, far from the rocks that make up the   (132)
mountains, much can be learned from the pebbles and boulders in the
river bars. Geologists in the early days, in particular, learned much
about the rocks of unknown mountains by examining the river gravels
derived from them.

The most prevalent rocks in the gravels of the Hoh are rounded, gray
pebbles of sandstone (fig. 6). Look at them closely to see the tiny,
closely knit sand grains. Some pebbles contain considerable white
mica (note 80), sparkling bright if turned to the sun just right. Some of
these gray sandstone pebbles are liberally sprinkled with flat ovoids
of black slate, chips of mud ripped off the ocean bottom when the
sand came sweeping in (fig. 9). A few gray pebbles of sand have a
layered, flaky structure, showing that the rock was subjected to much
squeezing during metamorphism (fig. 14).

Not so abundant as the sandstone pebbles but still common are
hard, black slate pebbles, most of which are flat (fig. 14). Some are
stained brown with limonite produced from breakdown of pyrite.
Break a brown-stained pebble open to see tiny cubes of silvery,
brassy pyrite. An exceptionally thorough search may turn up rounded,
hard greenstone, or altered basalt (fig. 6), carried from the Bailey
Range. We could correctly surmise that much of the mountain interior
was sandstone, with lesser amounts of slate, that some of the rocks
were metamorphosed, and that there were rare flows or dikes of
basalt.

Where the trail leaves the broad river valley to climb glacier-cut   (133)
benches to the Hoh River High Bridge (a mile or so upstream), the

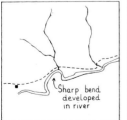 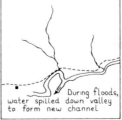 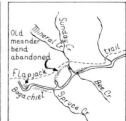

Fig. 85. Bogachiel River abandons a meander

hiker ascends a glacial step, now notched deeply by the river. When three major glaciers, one from the Hoh, one from the Glacier Creek drainage, and one from Falls Creek, all merged in the Hoh valley, their combined erosive force cut the main valley much deeper than the side valleys (notes 126, 176). Since that time the river has filed a slot in the glacial step.

The cutting power of a river can be fully appreciated from the center of the Hoh River High Bridge, which spans a deep slot in hard, black sandstone. At the bottom the flashing river grinds away to make the slot deeper.

(134)   The view after the final scramble to the top of either the moraine or the smooth rock bosses at the foot of the Blue Glacier leaves the hiker a bit awe-struck. For twenty miles he has been surrounded by forest and fern and here before him is an Ice Age world, nothing but snow, ice, and bare rock. In the immediate view is the Blue Glacier (fig. 86), one of the world's most thoroughly studied alpine glaciers (see Chap. 2, "Glaciers: The Heavier Hand"). The view from the moraine reveals many of the glacier's features. The details of its anatomy are more easily seen in late summer or early fall, when most of the previous winter's snow has melted back to reveal the underlying ice.

The glacier produces unique sounds, and the most dramatic is the thunder of falling ice. In the early days of man, the present Blue Glacier was farther advanced down-valley, and the roar of falling ice from its snout must have reverberated down the Hoh River valley. Reagan reported an Indian legend based on the roar of ice: "The Indians believe that in time of stormy weather a bird of monstrous size soars through the heavens and by the opening and shutting of his eyes it produces the lightning and by the flapping of its wings it produces the thunder and the mighty winds. This bird, they say, has its nest in a dark hole under the glacier at the foot of the Olympic glacial field and that its moving about in its home produces the 'thunder-noise' there."

(135)   The climber on the Blue Glacier or Snow Dome will see conspicuous layering in the rocks comprising Mounts Mathias and Mercury, east of the Blue Glacier. The sandstone beds, once lying flat in the ocean, represent floods of sand delivered to the ocean floor in a dense slurry. The slates are beds of mud and silt that accumulated more slowly. Mild currents carried suspended mud and silt into the deep water where it settled down to the bottom over many hundreds to thousands of years.

(136)   The clean rock on the east side of the Snow Dome affords a good look at ocean-bottom structures. The rocks are complexly folded and faulted, but a close look at any one rock face reveals thin sandstone

beds evenly alternating with silt beds. Some of the sandstone layers have graded bedding (note 123).

In spite of the clear preservation of the sedimentary structures, these rocks have recrystallized considerably and are rich in several metamorphic minerals. Most common are muscovite (white mica) and chlorite (green mica). Common also are prehnite and pumpellyite, generally in crystals too fine to see in hand specimens but visible under the microscope. If the assemblage of new minerals in the rocks is compared with similar minerals that the geochemist can create in the laboratory, where temperature and pressure are known, the geologist can estimate the depth and temperature at which the rocks formed. James Hawkins has estimated that the rocks reached 570°F. at a pressure of about 43,000 pounds per square inch. We can translate these numbers into more earthly terms if we consider how heavy rocks are and how hot the earth's interior. These characteristics of the earth indicate that the rocks were metamorphosed at no more than six miles below the surface and probably less if the rocks were squeezed by crustal movements at the same time (see Chap. 1, "Colliding Plates and Olympic Rocks").

Much of the Mount Olympus massif lies within a thick belt of highly disrupted rocks extending from High Divide to the Quinault River (note 148 and fig. 16). This belt is characterized by huge slabs of sandstone set in a slaty matrix (note 105). The sandstone making up Cat Peak and Mounts Mercury and Mathias is the largest slab (note 125). The peaks of Mount Olympus are smaller ones. These fossilized ocean-bottom sweepings can best be viewed from the nearly inaccessible upper South Fork of the Hoh River near the Hubert Glacier (fig. 87). (137)

One of the most easily recognized zones of disrupted rocks is exposed north of Mount Tom and extends southeast through the Paull Creek area and beyond (note 148). The zone is made of small blocks of sandstone, pieces of once-continuous beds, encased in a mass of slate. (138)

The barren lands below the White Glacier illustrate glacial features well. As the glacier melts, it releases the rock debris, *moraine,* that it carried (notes 134, 159). Where moraine has accumulated in thick piles near the glacier's end, it is called *terminal moraine;* near its side margins, *lateral moraine.* Sticking through the moraine are rock bosses polished by the glacier ice. Glacial outwash is morainal debris spread downstream by the glacial stream; it fills the broad valley in front of the glacier. (139)

As time passes, small ponds in the uneven, glaciated terrain will be filled by silt from the stream issuing from the snout of the glacier.

Fig. 86. Aerial views of Mount Olympus massif and sketch of the glacier's features. U.S.G.S. aerial photo GSWR 4-11

Alder brush will creep up the valley. Debris from adjacent steep hillsides begins to obliterate the glacial features.

(140)   Few hikers reach the inner sanctum of the uppermost Hoh River and the Hoh Glacier. Those who do may contemplate the dramatic story of a collapsing mountainside.

On the recently abandoned morainal flats below the glacier is an uneven, thick carpet of red-stained rock debris, obviously fallen from a fresh scarp on the east side of the valley. This landslide took place

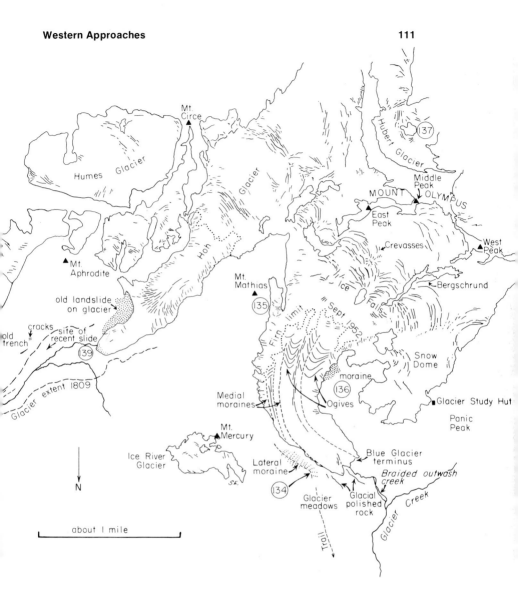

Mt.
Circe

Hubert Glacier

(37)

Humes Glacier

Glacier

Middle
Peak

MOUNT OLYMPUS

East
Peak

Crevasses

West
Peak

Hoh

Mt.
Aphrodite

Mt.
Mathias

(135)

Bergschrund

old landslide
on glacier

Firn limit

Sept. 1952

Ice Falls

Snow
Dome

old
trench

cracks

site of
recent slide

(139)

Glacier extent 1809

moraine

(136)

Medial
moraines

Ogives

Glacier Study Hut

Panic
Peak

N

about 1 mile

Mt.
Mercury

Ice River
Glacier

S.E.

Lateral
moraine

(134)

Blue Glacier
terminus

Braided outwash
creek

Glacier Creek

Glacier
meadows

Glacial
polished
rock

Trail

sometime after 1952, because aerial photographs (fig. 86) taken then do not show the slide. (An older slide is visible on the southeast half of the lower Hoh Glacier.) At the top of the disturbed scarp area is a trench, parallel to the scarp and similar to the ridgetop depressions described elsewhere (notes 109, 121), and above the trench are deep cracks on the ridge crest. These cracks show clearly in 1939 photographs, and at that time the Hoh Glacier terminus was just up-valley from the present scarp. Calvin Heusser showed that in 1809 the gla-

Fig. 87. Disrupted light-colored sandstone beds in sheared dark slate. View is from upper South Fork of the Hoh River at west side of Mount Olympus, with West Peak, second from left, and Middle Peak, far right. Note that vegetation has not been established on recently abandoned moraine in foreground.

cier terminus was almost a mile below the site of the slide. It seems possible that the cracks formed—the beginning collapse of the hill-side—as the glacier retreated, withdrawing its support from the valley side. Downriver from the present slide is at least one old trench, comparable to the one above the fresh scarp but now partly obscured by mature trees. As the glacier retreated, the oversteepened walls evidently collapsed into the valley.

South Fork of the Hoh River

(141)   From the South Fork Road (called the Mainline), where it stretches eastward across the gravel-terraced flats, look up to see Mount Tom. Its gently sloping north side is mantled by the White Glacier. Its precipitous south side rises 4,500 feet above the deep canyon of the South Fork of the Hoh.

(142)   The road to the beginning of the trail up the South Fork climbs up on a thick gravel terrace after it crosses the river. This is a remnant of

the valley-filling gravels prevalent all over the western Olympic Peninsula (see next note and Chap. 2, "Glaciers: The Heavier Hand").

Queets River

The Queets River road winds up a wide glaciated valley partly (143)
choked with gravels. The traveler will be unaware of the U shape of
the valley, but will be impressed by the flat terraces at different levels
traversed by the road. Hikers are particularly aware of river terraces
because they are level in a land of up and down. The terraces represent the place occupied by the river at some time past when it flowed
at a higher level. They may be composed of gravels deposited at a
time when a river could not carry all the debris delivered to it by side
streams or glaciers at its headwater. Various levels of terraces cut in
rock or gravel (fig. 88) are made by the river when its downcutting rate
is very slow, enabling it to meander sideways. Some terracelike features in Olympic valleys may have been cut by glaciers (notes 20, 30,
146).

Falls and rapids, such as Sams Rapids just north of the Queets (144)
campground, represent an interruption to the normal, less violent flow
of the river. The rapids formed when the river eroded into hard sandstone bedrock, not common in the gravel-blanketed Queets valley.
The rapids are not spectacular but probably hindered the early set-
tlers who traveled by canoe on the Queets.

The Kloochman Rock Trail climbs toward the top of a massive out- (145)
crop of black sandstone. Coal Creek, which drains the rock's southwest side, was probably named for traces of coaly plant material,

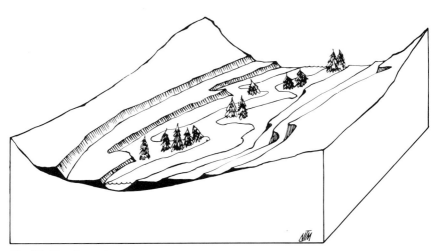

Fig. 88. River terraces

sparsely distributed in the sandstone of this area (see also note 50). The fragments of wood and plants accumulated along the beach of an ancient sea and then were washed into deeper water by sand-laden currents (note 135).

(146)   The hiker starting up the Tshletshy Creek Trail climbs out of the river bottom onto a broad gravel terrace; then he climbs again on terraces to the side of a bedrock knob above a deep gorge in Tshletshy Creek. Beyond this gorge the trail once again crosses a broad, terracelike feature before entering the steep-sided main canyon of Tshletshy Creek. The bedrock knob appears to be an isolated part of Sams Ridge to the south, cut off by Tshletshy Creek when it was diverted by glacial ice (fig. 89). The high flats traversed by the trail before entering the main canyon are the bottom of the old glaciated valley, which once bore the Tshletshy Creek glacier. The creek has cut an inner gorge into the old glaciated valley bottom and into glacial gravels in the Queets Valley.

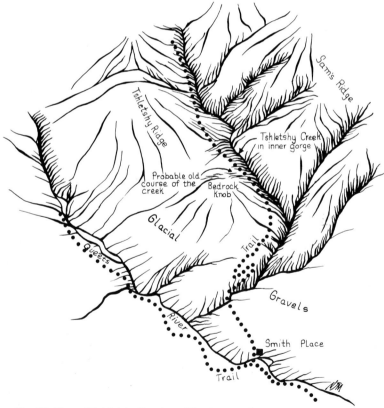

Fig. 89. View of Tshletshy Creek and its change of course

In its upper reaches, the Queets River has cut a deep canyon in (147) thick beds of black sandstone. Pelton Peak and the Valhallas to the northeast (fig. 90) and Tshletshy Ridge are rugged expressions of this resistant rock. It is folded and faulted in this area but nowhere as thoroughly disrupted as the rocks of Mount Olympus (note 137).

The ridge between Kilkelly Creek and Paull Creek, a seldom-visited (148) Elysian highland, spreads its heather meadows between deep gorges and rugged, pinnacled peaks. Its rounded form is produced by slate in a zone of breaking and crushing (notes 138, 154). Terrane to the east is also characterized by crushed and broken beds but of sandstone, which stand up in rugged peaks (note 137).

The upper Queets, like other large rivers in the Olympics, has cut a (149) 100- to 200-foot-deep inner gorge in the broad glaciated valley (note 86). In places below the Humes and Jeffers glaciers and at Service Falls, the gorge is almost vertical and inaccessible. It appears to be cut in glacial-carved steps of the valley (note 133).

Fig. 90. Thick beds of sandstone of the Valhallas above Geri-Freki Glacier

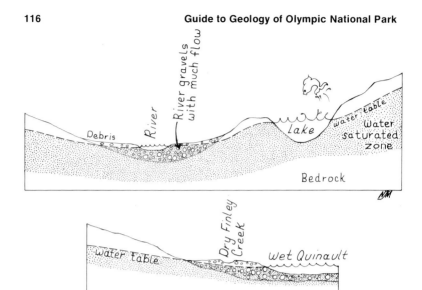

Fig. 91. Diagram showing how ground water reaches the surface in lakes and streams

North Fork of the Quinault River (North Side Road)

(150)   Finley Creek is a wide, bouldery creek bed dry most of the summer. A dry creek seems anomalous in the southwestern Olympics, where yearly rainfall exceeds 110 inches; but in fact Finley Creek is not really dry—the water is flowing deep in the thick gravels (fig. 91).

(151)   Irely Lake lies in a swamp in a deep, junglelike forest. The trail to it winds over bedrock knobs and across boggy depressions. This topography, a mixture of rock bosses and moraine, is the work of a glacier that came down the North Fork of the Quinault (see note 139 for description of similar but recently deglaciated terrain). The trail to Three Lakes follows Big Creek, which is completely lost in its own gravels and thick moraine southwest of Irely Lake (note 150) where some maps refer to the river's "underground passage."

(152)   Where the trail to Three Lakes crosses the main creek, gray beds of sandstone contrast sharply with black shales.

(153)   At Three Lakes, the traveler can see an advanced stage in the evolution of an alpine basin. The largest lake is surrounded by an encroaching layer of sphagnum moss. Eventually the lake will become a bog and then a meadow (note 120).

(154)   From Kimta Peak to Seattle Creek, the Skyline Trail traverses a belt of disrupted slate and thin-bedded sandstone (notes 138, 148, and fig. 16). Above Promise Creek, the traveler winds a tortuous route up, down, and through gullies cut in the contorted, broken rocks.

The North Fork Trail leaves the valley bottom gravels of Wolf Bar to (155) traverse various levels of terrace above the Quinault North Fork. This river, like others in the Olympics, has cut many miles of gorge (note 80) after glaciers smoothed its valley bottom.

At a riverside gravel bar (several miles above Francis Creek (156) Shelter) look up to spectacular sandstone cliffs on the north ridge of Mount Lawson. This sandstone appears to be a gigantic, coherent slab in a disrupted zone of slate (notes 105, 137).

The north ridge of Mount Lawson is an example of a double ridge- (157) top depression (fig. 92). The steep beds, unable to support them- selves against gravity, are peeling off the precipitous sides like the opening petals of a flower (notes 109, 121).

High on the west shoulder of Mount Christie is a parklike area of (158) small meadows and tarns. (To reach it, the stout of heart may leave the trail just downstream from the main creek draining the west side of Mount Christie and climb straight up the brushy mountainside.) A hanging glacier must have smoothed this parkland; its airy perch can be best appreciated by the view from the south rim overlooking a gully descending to Sixteenmile Shelter. A small stream gurgles peacefully over the meadows of the bench but plunges down the gully, where it is gnawing rapaciously into the high bench. In time, the flat of the bench will be changed to a steep canyonside.

Main Branch of the Quinault River (South Side Road)

Lake Quinault is dammed by moraine. When a glacier achieves a (159) delicate balance between melting at its lower end and snow accumu- lation at its upper end, its terminus tends to remain in one place. As the ice accumulating in the upper reaches flows down to melt at the terminus, rock debris that has fallen on the glacier or been plucked from its channel is carried along and accumulates at the end (note

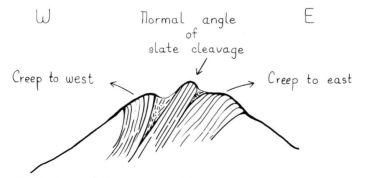

Fig. 92. Diagram of Mount Lawson ridge

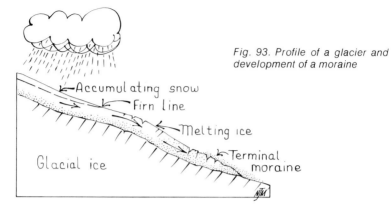

Fig. 93. Profile of a glacier and
development of a moraine

139). If the terminus stays in one place long enough, the heap of debris grows to become a terminal moraine (fig. 93).

(160)   The nature trail under the Willaby Creek Bridge (see the map in
U.S.F.S. Willaby Creek Forest Camp) passes above water-polished
bedrock outcrops of streaky white splinters and blocks of sandstone
in a black matrix of slate. This is one of the few readily accessible
views—albeit small-scale—of the disrupted rocks so common in the
core of the Olympics (notes 35, 105).

(161)   The glaciated U shape of the lower Quinault valley is not easily
appreciated from the valley bottom, but take a short side trip up the
Wright Canyon Road (turn right off the highway in Quinault 0.5 mile
north of the ranger station) to get a better impression of the steepness
of the U-shaped valley and to view a hanging side valley. The side
canyons were left hanging above the Quinault because their small
glaciers could not cut downward as fast as the larger Quinault Glacier.
At the upper end of Wright Canyon, clear-cut logging reveals the
U-shaped valley and the bowl-shaped cirque at the end.

(162)   From the farmlands northeast of Lake Quinault (about 2.7 miles
from Quinault Forest Service Ranger Station) the mountains on the
south side of the valley appear as squat, triangular hills. These
triangle-shaped peaks are not the main ridge, but spur ridges with
their ends truncated by the glacier that carved the valley (fig. 94).

(163)   A popular roadside stop is Merriman Falls (4.4 miles from the
Ranger Station). The falls tumble down the steep part of the U-shaped
valley from the small hanging valley of the Merriman Creek (note 161).
Almost every creek pouring into the Quinault valley must tumble down
this steep wall. Some of the bigger creeks upriver, like Howe Creek,
have cut a deep gorge in the steep valley side.

(164)   Just within Olympic National Park, where the Quinault Road swings
close to the river (10.6 miles beyond the Forest Service ranger sta-

tion), is a fine view up the U-shaped valley, of steep, forested mountains and snowy peaks beyond. The valley was once filled with several thousand feet of glacial ice. Also at this point is a conspicuous 15-foot boulder of metamorphosed basalt between the river and the road. Since there is no such rock exposed here in the valley, it is probably a glacial erratic carried to this place by the glacier, conceivably from basalt peaks at the headwaters of Cannings Creek, which joins the Quinault just up the road.

At the second conspicuous opening to the river where the road (165) hugs cliffs above river bars and rushing water (15.1 miles from the ranger station), look for good exposures of inner Olympic slates. The slate here breaks into large, elongate sticks owing to the forces that have squeezed the slate (note 64). In the slate are pods of sandstone that appear to have been torn off sandstone beds during the deformation (note 35).

At Graves Creek, the trail along the old road leaves the Quinault (166) River bottom, crosses a bridge high above the Graves Creek gorge,

Truncated spurs are just preamble to the main ridge crest

Lake Quinault

Fig. 94. Scene along Quinault River road, showing truncated mountain spurs

and ascends a peculiar small valley carrying only a trickle of water. The river that once flowed here is now cutting a deep canyon just over the ribs of rock to the northwest of the road. Of several possible histories, one is that the small, dry valley followed by the road was probably part of the original river valley before the last ice advance. At some point in the glacial retreat, the main river cut into rock on the western side of the valley—perhaps the ice lingered here on the east side forcing the river to the west—and by the time the ice was gone, the river had notched itself into the valley side and could not return to its earlier channel (fig. 95). Uplift (see Chap. 2, "Rising Land and the Olympic Scene") of the Olympics has increased the cutting so the river is now well below the old valley floor.

(167)   Near Pony Bridge and in the gorge beneath it are exposures of disrupted slate and sandstone. The broken and torn nature of the rock can best be seen on water-polished outcrops just downriver from the bridge. Slivers and fragments of sandstone beds are surrounded by the slaty matrix (notes 35, 105).

The gorge of the Quinault at Pony Bridge has been cut into an old valley bottom or bottoms, represented by terraces at various levels that are traversed by the trail (note 80).

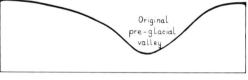

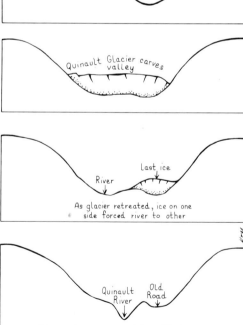

Fig. 95. Development of Quinault gorge above Graves Creek

Mary Ann Lakes, seldom visited but worth the effort for experienced   (168)
hikers, are beautiful examples of glacial tarns. The snow lingers long
into the summer in the uppermost lake basin. The lower lake is slowly
filling in, and the sphagnum moss has reduced its extent considerably.
In a basin below the main lake, a meadow is disappearing. Below this
meadow are small boggy clearings in tall trees, the only remaining
evidence of the glacial bareness that once ruled (notes 77, 120).

The Quinault River Trail wanders along the valley bottom over a va-   (169)
riety of valley materials—sods, debris, terraces, river gravels, and
moraine. At Noname Creek, Pyrites Creek, and Lamata Creek the trail
swings onto prominent *alluvial fans,* the cone-shaped deposits of
gravel dropped by these swift creeks when they reach the flat floor of
the valley and no longer have the energy to carry heavy loads of debris.

The footbridge at the lower end of Enchanted Valley crosses   (170)
bouldery rapids; the river and the trail wind around large boulders just
northeast of the bridge. This debris appears to be a terminal moraine
of the Quinault Glacier where it paused long enough in its retreat to
pile up its conveyor-belt load (note 159).

The flat floor of the Enchanted Valley is a striking feature in a steep   (171)
land. The valley, of course, was filled at one time by the Quinault Glacier, and glaciers can carve flat floors. Here, however, the flat floor
gives evidence of a former lake, dammed by the moraine at the trail
crossing (note 170). The lake may have been very short-lived, but the
moraine dam held the infilling debris and forced the river to spread
the gravels evenly from valley side to valley side (fig. 96). Ultimately

*Fig. 96. The U shape of the Enchanted Valley was carved by a glacier. View is
northeast to Mount Anderson*

Fig. 97. Views of the Anderson Glacier showing its retreat between 1927 and 1965, looking north fro

he morainal viewpoint, and with the base of the East Peak of Mount Anderson right. Photos by W. M. Cady

the flat floor of the valley will be removed by the river as it cuts through the dam. Perhaps a few flat terraces will remain to show future travelers that a lake was once here.

Another unusual aspect of the valley is its openness. At this elevation (about 2,000 feet) we would expect heavy timber. However, the persistence of snow cones at the base of the western wall of the valley suggests that, at least until recently, avalanche snow so filled the valley in the wintertime that conifers could not get established. Now the brush is so thick that the conifers still have not encroached much. The existence of the chalet shelter, built in 1930, suggests that today avalanches do not play so important a part in keeping the valley free of trees.

The snow cones themselves are commonly full of rock debris, delivered to the valley bottom by winter avalanches from the steep east face of Chimney Peak. These avalanches are thus important agents of erosion as the falling snow and rock cut steep chutes in the mountain face.

(172)   The green, terraced cliffs forming the 5,000-foot east face of Chimney Peak are composed of torn and smashed slate and minor sandstone. This belt of disrupted slates stretches from the North Fork of the Skokomish to Olympic Hot Springs (fig. 16). All along it and especially along the west side of Enchanted Valley, the observer can find evidence of crushing, grinding, bending, and breaking of the rocks (see Chap. 1, "Colliding Plates and Olympic Rocks," and other examples, notes 35, 105). On the cliffs below Chimney Peak and on the peak itself are small and large pods of basalt lava, crushed and metamorphosed to greenstone. These pieces of hard rock may have been dragged into the softer shales from more extensive volcanic layers by the milling of the rock under the continent's edge (fig. 24). Also, abundant in these rocks are veins of quartz with well-developed quartz crystals up to several inches long (note 100). Look for crystals weathered from the rock in the debris piled up at the base of the cliffs. Inspect the crystals, but leave them for others to look at.

(173)   Where the trail zigzags up to Anderson Pass, light-reddish-brown sandstone crops out. The sandstone beds are also well exposed near the falls below the Anderson Glacier, visible from the trail. The dark rust color on some surfaces of this rock is caused by the hydrous oxide of iron, limonite, derived from iron-bearing minerals in the rock.

(174)   An unmistakable moraine crowns the hill overlooking the Anderson Glacier (near the end of the way trail from Anderson Pass). The sharp-ridged pile of debris is almost barren, although the alpine wildflowers and heather have made a toe hold in its loose rubble. The view of the glacier sweeps over a basin filled with even more barren

moraine and glacier-smoothed rock knobs. The glacier has retreated considerably since man first began to visit it, and the silt-laden lake is forming where ice used to be (fig. 97). The Anderson Glacier and an unnamed glacier over the ridge to the west are but small symbols of the ice that once stretched from Mount Anderson down the valley almost to the ocean.

Southern Approaches

Wynoochee River

(175)   Near Wynoochee Falls (U.S.F.S.) Campground, the U-shaped profile of the Wynoochee valley is readily apparent. The valley glacier here was about 3,000 feet thick, and it ground back the valley walls to their present steepness. The creeping river of ice extended 25 miles down the valley, where it spread a broad piedmont tongue on the flats south of the mountain front. Just above the campground, Wynoochee Falls are cut in bedrock on the riser of a glacial step (note 133).

(176)   The West Branch of the Wynoochee is a hanging tributary; its smaller glacier could not cut down as fast as did the ice in the main Wynoochee. And the North Fork of the West Branch is a hanging tributary to the West Branch. Observe the narrow rock slot cut by the North Fork in the valley wall. Rivers cut downward by abrasion; processes like slump, landslide, and washing by rain widen the sawlike cut into the typical V-shaped valley (fig. 98). Here the widening processes have been slower than downcutting and have not had time to change the saw cut to a V shape.

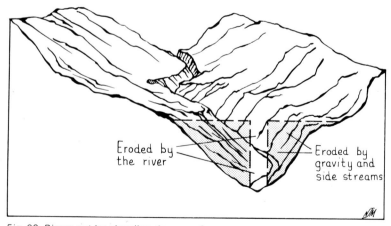

Fig. 98. Rivers cut by abrading downward

To gain a spectacular view of the steep walls of the glacier-cut   (177)
West Branch of the Wynoochee, travel the high logging road leading
to the upper west side of the North Fork. Also seen from this road
are outcrops of basalt, the Crescent Formation, and red limestones of
the Olympic basaltic horseshoe (see note 26).

A western branch road (F.S. 2363) from the Forest Service road to   (178)
the Wynoochee Pass Trail brings the traveler to a small, subalpine
basin carved by a glacier from thick beds of gray sandstone. The
basin, partly logged, once bore small heather meadows and
sphagnum-ringed ponds nestled on forested benches (see note 120)
that rise up to the south and merge in the sharp summits of Three
Peaks. To the north, Discovery Peak and ridges to the west are carved
from basalt slivers of the inner Olympics (fig. 16).

South Fork of the Skokomish River

Just before Church Creek on the road up the South Fork of the Sko-   (179)
komish (F.S. 239; 13.4 miles from Fir Creek Forest Service Guard Station), look up to see the steep basaltic cliffs of Mount Church. The
route up the South Fork is mostly in basalt of the Olympic horseshoe
(Crescent Formation). Although the wide outcrop pattern of basalt on
the south side of the range would lead us to think the basalt pile was
as thick here as it is on the east side of the Olympics (note 26 and fig.
16), this does not appear to be so. On the east side the flows stand
vertically, and we can measure their true thickness on edge. Here they
are folded, and their true thickness is unknown because we do not
see them on edge.

Church Creek (14.6 miles from Fir Creek Guard Station) has cut a   (180)
deep, narrow gorge where it spills out of its hanging valley into the
deeper, glacier-cut South Fork valley (note 161).

The new trail to Sundown Pass skirts a deep basalt gorge cut by the   (181)
South Fork. The river enters this gorge abruptly in a remarkably
narrow slot from a wider valley near the junction of Rule Creek. If this
wider valley were carved from soft sedimentary rocks, we might consider that the river had thrown its energy into widening the valley in
soft rock while it cut its way straight down the harder basalt (note
176). But the valley immediately above the slot is also cut in basalt,
suggesting that the wider valley occupies a basin that was scooped
out by the South Fork Glacier. When the glacier melted, the river was
dammed by a rock buttress until it succeeded in cutting the gorge.
Thick gravel deposits along the trail just above the river crossing may
be material dumped into the temporary lake above this rock dam.

Where the trail climbs out of the river bottom are some large basalt   (182)
boulders, either glacial erratics or blocks fallen from basalt cliffs on
the east.

(183)   Beyond Startup Creek, the trail climbs abruptly up a glacier-carved step (note 126) underlain partly by basalt. The river descends this step in a narrow gorge to the east.

(184)   The small boggy meadows on benches of the upper cirque above the step are probably filled glacier tarns. The south side cirque will soon (geologically speaking) be forested, and only the steplike topography will remain as a signature of the icy sculptor (see notes 77, 120). The relicts of the glacial past are much fresher on the north side of Sundown Pass, where glacier-carved cirques are still barren and tarns such as Lake Sundown still exist.

Glossary

The terms included in this glossary are used frequently throughout the book. A reference in parentheses after a definition indicates a place or places in the text where a term is defined more completely.

Basalt. A dark, dense volcanic rock generally composed of feldspar and varying amounts of rock glass, iron oxides, pyroxene, and other iron-magnesium silicates (Chap. 1; fig. 6; note 46). *Pillow basalt* forms when lava erupts into water and forms pillowlike globular masses (Chap. 1; note 25).

Bed, Bedding. A layer in rock that represents deposition of relatively uniform material, such as sand, which sets the bed off from adjoining beds of different material such as mud or silt. Most beds are deposited in a nearly horizontal plane (Chap. 1; notes 55, 135).

Boss. A rounded rock knob making a small bump or hillock.

Breccia. A rock consisting of angular fragments cemented together. Contrasts with conglomerate, which has rounded fragments. *Volcanic breccia* is predominantly volcanic rock fragments. *Sedimentary breccia* is composed of sedimentary rock fragments such as in shale (slate)-chip breccia.

Calcite. A mineral composed of calcium carbonate. It is generally white, can be scratched with a knife, and is the predominant constituent of limestone.

Cirque. Bowl-shaped recess at the head of a glacier-carved valley; the starting place of glaciers (Chap. 2).

Clay minerals. A family of complex aluminum silicates with much water and potassium, sodium, calcium, magnesium, and iron. Many are too small to see except with an electron microscope. They are the characteristic end product of rock weathering and are concentrated in soils.

Conglomerate. A rock consisting of rounded fragments cemented together. Conglomerates are derived from gravel (Chap. 1; fig. 6).

Creep. The slow movement of rock debris downhill under the pull of gravity (notes 11, 71).

Crust (of the earth). Outer layer of the earth. It consists of lighter material than the inner layers and core of the earth. The *continental crust* is mostly granite and granitelike rocks; *oceanic crust* is mostly basalt and related rocks.

Diabase. A granular igneous rock generally formed from a melt at relatively shallow depths in the crust. Composed of calcium feldspar and pyroxene plus iron oxides, it is similar to basalt but has larger, visible crystals (Chap. 1; note 46).

Differential erosion. Process whereby soft rocks are eroded faster than hard rocks. Generally, the hard materials stand higher or jut out from the softer materials (Chap. 2; notes 17, 55).

Epidote. A mineral generally formed by metamorphism. It is a complex aluminum silicate of calcium, iron, and water, and is apple green in color (note 2).

Fault. Break between two blocks of rock that have moved relative to each other and parallel to the break. A *fault zone* is a band of subparallel faults. Rock in a fault zone is usually highly broken and crushed. Faults range from a few inches to many miles long and fault zones from less than an inch to several miles wide (Chap. 1; notes 18, 32).

Feldspar. General name for a mineral group of complex aluminum silicates of potassium, calcium, and sodium. Generally white or pink, blocky, and cannot be scratched with a knife. One of the most common rock-forming minerals (Chap. 1).

Fold. A bend or curve in any layer, usually a bed, of rock. If the folded bed is bent around older beds (fig. 19), it is called an *anticline;* if the bed is bent around younger beds, a *syncline.*

Gabbro. A granular igneous rock composed of calcium feldspar and pyroxene plus iron oxides. It is like a diabase but generally formed at greater depth and coarser grained (Chap. 1; note 46).

Glacial step. A steplike drop in a valley caused by local quarrying of the bottom by a glacier (notes 30, 133).

Granite. A granular igneous rock composed of feldspar, quartz, and iron-magnesium silicates such as hornblende and biotite. It crystallizes deep in the earth from a melt and is the characteristic rock of many mountain ranges. Usually light in color but sprinkled with dark minerals.

Greenstone. A dense, green, somewhat featureless rock containing epidote, sodium feldspar, and chlorite. It is derived from basalt, diabase, or gabbro by metamorphism (Chap. 1).

Hanging valley. A tributary valley with a higher floor than that of the trunk valley that it joins. The trunk valley is carved deeper because it had a larger glacier than the tributary valley (notes 163, 176).

Hematite. A mineral composed of iron oxide. Derived by oxidation of other iron-bearing minerals, it constitutes the red color in many rocks (note 2).

Hornblende. A common rock-forming mineral, colored dark green to black and growing in elongate prisms. A complex aluminum silicate of calcium, iron, magnesium, and water, it cannot be scratched with a knife.

Joints. Cracks in rocks, generally occurring in a uniform, parallel pattern. In many rocks, joints are thought to form when the rock expands as erosion removes the overlying, once confining material (note 8).

Lava. Molten rock on the earth's surface. Sometimes used casually to refer to the rock of a cooled and hardened lava flow (Chap. 1).

Limestone. A rock consisting mainly of calcite, commonly derived from the skeletons of marine animals. It may be white, gray, black, or red; fine-grained and smooth to coarsely crystalline; easily scratched with a knife (Chap. 1).

Limonite. A mineral composed of iron oxide plus water. Derived from weathering of iron-bearing minerals, it is the predominant constituent of rust and is commonly the brown color found in rocks.

Metamorphism. In rocks, the recrystallization of minerals to new ones that are stable under changed conditions of temperature and pressure, and/or solutions promoting chemical reaction. A rock so changed is a *metamorphic rock* (Chap. 1; note 136).

Mica. A mineral group of complex aluminum silicates of potassium, aluminum, magnesium, and iron with water. Consisting of thin, glassy, and elastic flakes, mica can be scratched with a knife. White mica is called *muscovite;* black mica (less abundant in the Olympics) is called *biotite;* green mica (which is nonelastic) is called *chlorite.*

Mineral. A naturally occurring chemical compound with a definite range of physical and chemical characteristics. Minerals in various proportions make up rocks. Significant minerals in the Olympic Mountains are calcite, clay, epidote, feldspar, hematite, hornblende, limonite, mica, pumpellyite, pyrite, pyroxene, quartz, and zeolite (see individual definitions).

Moraine. Unsorted rock rubble, sand, silt, and clay deposited by a glacier (Chap. 2).

Phyllite. A fine-grained metamorphic rock composed of quartz, feldspar, chlorite, and other micas; derived by recrystallization from shale. It tends to break into thin folia that are shiny from minute, aligned crystal faces of mica (Chap. 1; fig. 14).

Pumpellyite. A mineral derived from other minerals by metamorphism and rarely seen without a microscope in Olympic rocks. It is a complex aluminum silicate of calcium with water and is green in color (note 136).

Pyrite. A whitish yellow, shiny metallic mineral consisting of iron sulfide. It commonly occurs as tiny cubes. Small amounts are found in many different rocks; it breaks down easily to hematite and limonite.

Pyroxene. A mineral occurring as stubby, black crystals that cannot be scratched with a knife. It is a complex silicate of calcium, iron, and magnesium. Hard to see, but common in Olympic basalt, diabase, and gabbro (Chap. 1; fig. 6; note 46).

Quartz. Most common rock-forming mineral and easily recognized in many Olympic rocks. Composed of silicon dioxide, it is clear to white, glassy, and grows, where unobstructed, in pointed crystal prisms (Pt. I; note 100).

Rock. Rocks are the fundamental substance of the earth and are themselves made of one or more minerals (q.v.). Common rocks of the Olympics are basalt, breccia, conglomerate, diabase, gabbro, granite, greenstone, limestone, phyllite, sandstone, semischist, shale, and slate (see individual definitions).

Sandstone. A sedimentary rock made of sand grains eroded from pre-existing rocks (Chap. 1; fig. 6).

Semischist. A streaky, layered metamorphic rock with the tendency to break on thin planes; derived from sandstone during strong deformation and some recrystallization (Chap. 1; fig. 14).

Shale. A fine-grained sedimentary rock derived from mud; commonly finely bedded with a tendency to break along bedding planes (Chap. 1; fig. 6).

Slate. A fine-grained metamorphic rock characterized by its tendency to break along minutely spaced parallel cracks (cleavages). Generally dull gray or black, it is derived from shale; with further recrystallization it becomes phyllite (Chap. 1; fig. 14; note 64).

Talus. Slope of coarse rock rubble that accumulates at the base of a cliff. The rubble is generally coarser at the bottom than at the top of the slope.

Terrace (river or glacial). A relatively flat-topped surface above a river, usually much longer than wide and running parallel to the valley side. It represents a higher level of river or glacier action. Terraces can either be cut in bedrock or made of gravel and sand deposited by the river.

Zeolite. A mineral family of complex aluminum silicates of sodium, calcium, potassium, and water. They are white, hard to soft, fibrous to blocky, and hard to recognize (Chap. 1).

Reading and References

General Geology

Cloos, Hans. *Conversation with the Earth.* New York: Alfred A. Knopf, 1953. A geologist-poet tells how it is to be a geologist and how he investigated some of the earth's puzzles.

Cox, Allan. "Geomagnetic Reversals." *Science* 163 (1969): 237-54. Technical; cornerstone of the plate tectonic theory.

Leveson, David. *A Sense of the Earth.* New York: Natural History Press, 1971. Another geologist-poet is inspired by new geologic discoveries, including man's fundamental relationship to his planet.

McKee, Bates. *Cascadia: The Geologic Evolution of the Pacific Northwest.* New York: McGraw-Hill Book Co., 1972. An excellent summary of geologic principles and places; for the student and geology buff alike.

Raft, Arthur, and Ronald G. Mason. "Magnetic Survey off the West Coast of North America, 40° N. Latitude to 52° N. Latitude." Geological Society of America *Bulletin* 22 (1961): 1267-70. Technical but important documentation supporting plate tectonics.

Shelton, John S. *Geology Illustrated.* San Francisco: W. H. Freeman and Co., 1966. An elementary text with magnificent photographs.

Takeuchi, H., S. Veyeda, and H. Kanamori. *Debate about the Earth: An Approach to Geophysics through Analysis of Continental Drift.* Rev. ed. San Francisco: Freeman, Cooper and Co., 1970. Excellent and stimulating layman's introduction to geophysics.

Olympic Geology

Arnold, Ralph. "Notes on Some Rocks from the Sawtooth Range of the Olympic Mountains, Washington." *American Journal of Science* 28 (1909): 9-14. Technical but interesting historical lore about Black and White mines.

Bretz, J H. *Glaciation of the Puget Sound Region.* Washington Geological Survey Bulletin 8. Olympia, Wash.: Washington State Geological Survey, 1913. Earliest account of the Pleistocene scene.

Crandell, Dwight R. "The Glacial History of Western Washington and Oregon." In *The Quaternary of the United States,* edited by H. E. Wright, Jr., and D. G. Frey, pp. 341-53. (A review volume for the Seventh Congress of the International Association for Quaternary Research.) Princeton, N.J.: Princeton University Press, 1965. Technical.

Danner, W. R. *Geology of Olympic National Park.* Seattle: University of Washington Press, 1955. For the general reader.

Hawkins, James W., Jr. "Prehnite-Pumpellyite Facies Metamorphism of a Graywacke-Shale Series, Mount Olympus, Washington." *American Journal of Science* 265 (1967): 798-818. Technical.

Heusser, Calvin J. "Palynology of Four Bog Sections from the Western Olympic Peninsula, Washington." *Ecology* 45, no. 1 (1964): 23-40. Technical account of plant succession deduced from fossil pollen.

Park, Charles F., Jr. *Manganese Resources of the Olympic Peninsula, Washington: A Preliminary Report.* U.S. Geological Survey Bulletin 931-R, pp. 435-57. Washington, D.C.: U.S. Government Printing Office, 1942. Technical description of some Olympic manganese prospects.

————. "Structure in the Volcanic Rocks of the Olympic Peninsula, Washington." Geological Society of America *Bulletin* 61, no. 12 (1950): 1529. Abstract of technical talk on Olympic structure. First published insight of structure as we see it today.

Rau, W. R. *Foraminifera from the Northern Olympic Peninsula, Washington.* U.S. Geological Survey Professional Paper 374-G, pp. G1-G33. Washington, D.C.: U.S. Government Printing Office, 1964. Technical.

————. *Geology of the Washington Coast between Point Grenville and the Hoh River.* Washington Department of Natural Resources, Geology and Earth Resources Division Bulletin 66. Olympia, Wash.: Washington Department of Natural Resources, 1973. A good companion for the beach hiker.

Reagan, Albert B. "Some Notes on the Olympic Peninsula, Washington." Kansas Academy of Science *Transactions* 22 (1909): 131-238. A great deal of somewhat quaint natural science, including early views of the geology, plus many notes on Indian legends.

Sharp, R. P. *Glaciers.* Eugene: University of Oregon Books, 1960. Excellent, understandable discussion of glacial features and mechanisms, with emphasis on the Blue Glacier.

Snavely, P. D., Jr., and H. C. Wagner. *Tertiary Geologic History of Western Oregon and Washington.* Washington Division of Mines and Geology Report of Investigations 22. Olympia, Wash: Washington Department of Natural Resources, 1963. Excellent, simplified treatment, with clear diagrams.

Tabor, Rowland W. *Geologic Guide to the Deer Park Area, Olympic National Park.* Port Angeles, Wash.: Olympic Natural History Association, 1965. Nontechnical.

————. *Geologic Guide to the Hurricane Ridge Area.* Port Angeles, Wash.. Olympic Natural History Association, 1969. Nontechnical.

————. "Origin of Ridge Top Depressions by Large-scale Creep in the Olympic Mountains, Washington." Geological Society of America *Bulletin* 82 (1971): 1811-22. Technical.

Weaver, Charles E. *Tertiary Stratigraphy of Western Washington and Northwestern Oregon.* University of Washington Publications in Geology 4. Seattle: University of Washington Press, 1937. Technical and historical.

Olympic Geologic Maps

Brown, R. D., Jr., H. D. Gower, and P. D. Snavely, Jr. *Geology of the Port Angeles–Lake Crescent Area, Clallam County, Washington.* U.S. Geological Survey Oil and Gas Investigations Map OM 203 (1960). Detailed map.

Cady, W. M., M. L. Sorensen, and N. S. MacLeod. *Geology of The Brothers Quadrangle, Washington.* U.S. Geological Survey Geological Quadrangle Map GQ-969, scale 1:62,500 (1972). Detailed colored map for the real geology buff.

Cady, W. M., R. W. Tabor, N. S. MacLeod, and M. L. Sorensen. *Geology of the Tyler Peak Quadrangle, Washington.* U.S. Geological Survey Geological Quadrangle Map GQ-970, scale 1:62,500 (1972). Detailed colored map.

Gower, H. D. *Geologic Map of the Pysht Quadrangle, Washington.* U.S. Geological Survey Geological Quadrangle Map GQ-129, scale 1:62,500 (1960). Detailed colored map.

Rau, W. W. *Geology of the Wynoochee Valley Quadrangle.* Washington Department of Resources Bulletin 56. Olympia, Wash.: Washington Department of Natural Resources, 1957. Technical.

————. Stratigraphy and Foraminifera of the Satsop River Area, Southern Olympic Peninsula, Washington. Washington Division of Mines and Geology Bulletin 53. Olympia, Wash.: Washington Department of Natural Resources, 1956. Technical.

Stewart, R. J. "Petrology, Metamorphism and Structural Relations of Graywackes in the Western Olympic Peninsula, Washington." Ph.D. dissertation, Stanford University, 1970. Technical.

Tabor, R. W., R. S. Yeats, and M. L. Sorensen. *Geologic Map of the Mount Angeles Quadrangle, Washington.* U.S. Geological Survey

Geological Quadrangle Map GQ-958, scale 1:62,500 (1972). Detailed colored map.

General Olympic Lore

Gilman, S. C. "The Olympic Country." *National Geographic Magazine* 7 (1896):133-40. Interesting historic account.

Olympic Mountains Rescue. *Climber's Guide to the Olympic Mountains.* Seattle, Wash.: The Mountaineers, 1972. A must for the off-trail hiker.

O'Neil, Lt. Joseph. *Report of the Exploration of the Olympic Mountains, Washington, from June to October 1890.* Senate Document no. 59, 54th Congress, 1st Session. Washington, D.C., 1896. Fascinating, understated account of early exploration.

Webster, E. B. *Fishing in the Olympics.* Port Angeles, Wash.: The Evening News, Inc., 1923. Quaint and historic.

————. *The Friendly Mountain.* Port Angeles, Wash.: The Evening News, Inc., 1921. Charming and personal old-fashioned essay about Mount Angeles—its terrain, flora, fauna, and friends in the early days.

Wood, Robert L. *Across the Olympic Mountains: The Press Expedition, 1889–90.* Seattle: University of Washington Press, 1967. A delightful account, much in the words of the explorers themselves (see Part I).

————. *Trail Country: Olympic National Park.* Seattle, Wash.: The Mountaineers, 1968. The most complete Olympic trail guide available. A wealth of other Olympic lore.

Index